IONICS

A SCIENCE OF AWARENESS

IONICS INSTITUTE

IONICS

A Science of Awareness

Mee, J.L.
Ionics – A Science of Awareness

ISBN: 978-1-892654-47-2

1. Consciousness 2. Spirituality
3. Neuroscience

Published in the United States of America

Brightwaters Books
brightwatersbooks.com

Foreword

In an era when science seeks to quantify even the most elusive aspects of human experience, Ionics ventures boldly into the uncharted territories of consciousness. This work introduces a pioneering mathematical framework that captures awareness as a measurable phenomenon, bridging the domains of neurology, quantum mechanics, and philosophy. Through a series of axioms and equations, Ionics posits that awareness is not merely an emergent property of neural networks but a fundamental element of the universe itself—a force as tangible and quantifiable as any found in physics.

This text invites readers to consider consciousness through a new lens: as packets of *iotons*—awareness composed of quantifiable units capable of interaction, transformation, and integration with physical systems. Within this model, awareness oscillates between conscious and unconscious states, governed by laws of conservation and dual states of form and formlessness, a paradigm inspired by quantum theories yet fully attuned to the nature of human cognition.

The book challenges established scientific principles, proposing genetic and neurological innovations for cognitive enhancement, mood regulation, and memory manipulation. It not only ventures into philosophical queries about the origins of consciousness but also presents actionable insights for expanding human potential, grounded in genetics and leading-edge neuroscientific research.

For scientists, philosophers, and seekers of knowledge alike, Ionics provides a revolutionary path forward—a bridge between the abstract and the measurable, urging us to rethink the nature of mind, memory, and reality. It is a call to embrace a new science of consciousness, one that respects both the complexity of our inner lives and the rigor of scientific inquiry.

Introduction

The pursuit of understanding consciousness has long been the domain of philosophy, psychology, and, more recently, neuroscience. Yet, despite decades of research and exploration, the phenomenon of awareness has remained an enigma—a frontier as vast and mysterious as the cosmos itself. Ionics represents a groundbreaking effort to cross this frontier, offering a scientific approach to consciousness that integrates mathematics, neurology, and principles of quantum physics into a cohesive framework.

At the heart of Ionics lies a bold proposition: that awareness can be quantified, modeled, and ultimately optimized. This work introduces a set of axioms and formulas that express consciousness as a dual state of free and applied awareness, mediated through quantifiable units called "iotons." In this model, awareness flows through both conscious and unconscious channels, interacting dynamically with the brain's neural architecture in ways that science can measure, modify, and harness.

The implications of this approach are far-reaching. By understanding consciousness as a granular phenomenon governed by scientific laws, we can begin to consider practical applications for enhancing human potential.

Techniques for modulating brainwave frequencies, genetically modifying neural pathways, and selectively targeting memory could lead to advances in cognitive enhancement, emotional resilience, and even mental health treatment. For those who explore mindfulness, meditation, or spiritual practices, Ionics offers a scientific bridge to understanding how these states arise and how they might be cultivated with precision.

Throughout this text, readers will encounter foundational concepts that redefine awareness, memory, and perception in quantitative terms. The book explores the mind-matter interface, drawing from established scientific principles to suggest how consciousness and physical reality interconnect. Each chapter invites you to rethink what it means to be aware, presenting a framework that is both visionary in scope and rooted in rigorous scientific inquiry.

As you embark on this exploration, approach it with an open mind. Ionics does not merely seek to describe consciousness but to transform how we understand and engage with it. It is an invitation to explore the boundaries of awareness, to question, and ultimately to consider that consciousness itself may be as foundational to the universe as energy or matter. This is more than a study; it is a blueprint for the evolution of human cognition and self-awareness. Welcome to the future of consciousness science.

IONICS
A SCIENCE OF AWARENESS

IONICS

Ionics is a new cognitive science which enables the development of genomic technologies for accelerating human consciousness evolution.

GENETIC COGNITIVE ENHANCEMENT

Gene editing has been successfully deployed to enhance cognition in laboratory animals, boosting memory by 100%.

CRISPR was used to modify neuron electrodynamics to improve the animals' attention, enabling them to form stronger memories.

Attention in mice is a proxy for awareness in humans.

Thus, we may be at the threshold of creating new states of consciousness.

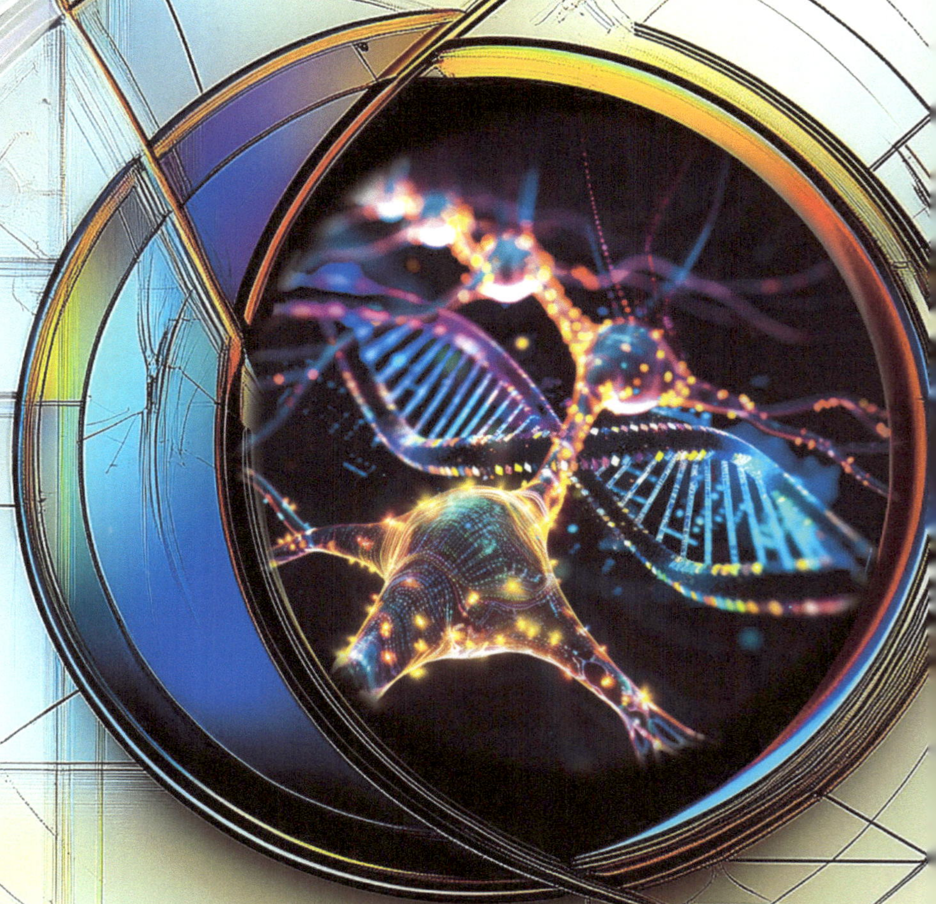

PERSPECTIVE

The experimental CRISPR design was based on a new understanding of the mind/matter interface, which is presented here for the first time.

The nature of this interface has eluded philosophers and scientists alike for centuries.

Historically, when major aspects of the universe have been captured in equations, it has led to paradigm shifts in science.

Codifying the behavior of the mind/matter interface in formulas would represent a

major milestone in both philosophy and science, with far-reaching implications.

9 KEY INSIGHTS

1. Visualizing awareness in packets allows us to treat it numerically, yielding a mathematics of consciousness.

2. Meditation and psychedelics reveal larger dimensions of human awareness.

3. This latent awareness is encapsulated in active unconscious memories which are often superfluous.

4. Dormant awareness can be restored back to the individual by retrieving it from inessential memories.

5. A human being's unconscious memory field is joined in an electrical circuit with their body's neurology.

6. Brainwaves are the physical manifestation of active unconscious memories.

7. Unnecessary memories can be disrupted by altering neuron timing in the neural pathways their brainwaves follow.

8. Disrupted memories lose their integrity, causing their information to collapse and releasing their enclosed awareness.

9. Modifying neuron timing in certain neural networks can convert nonessential brainwave activity into enriched conscious awareness.

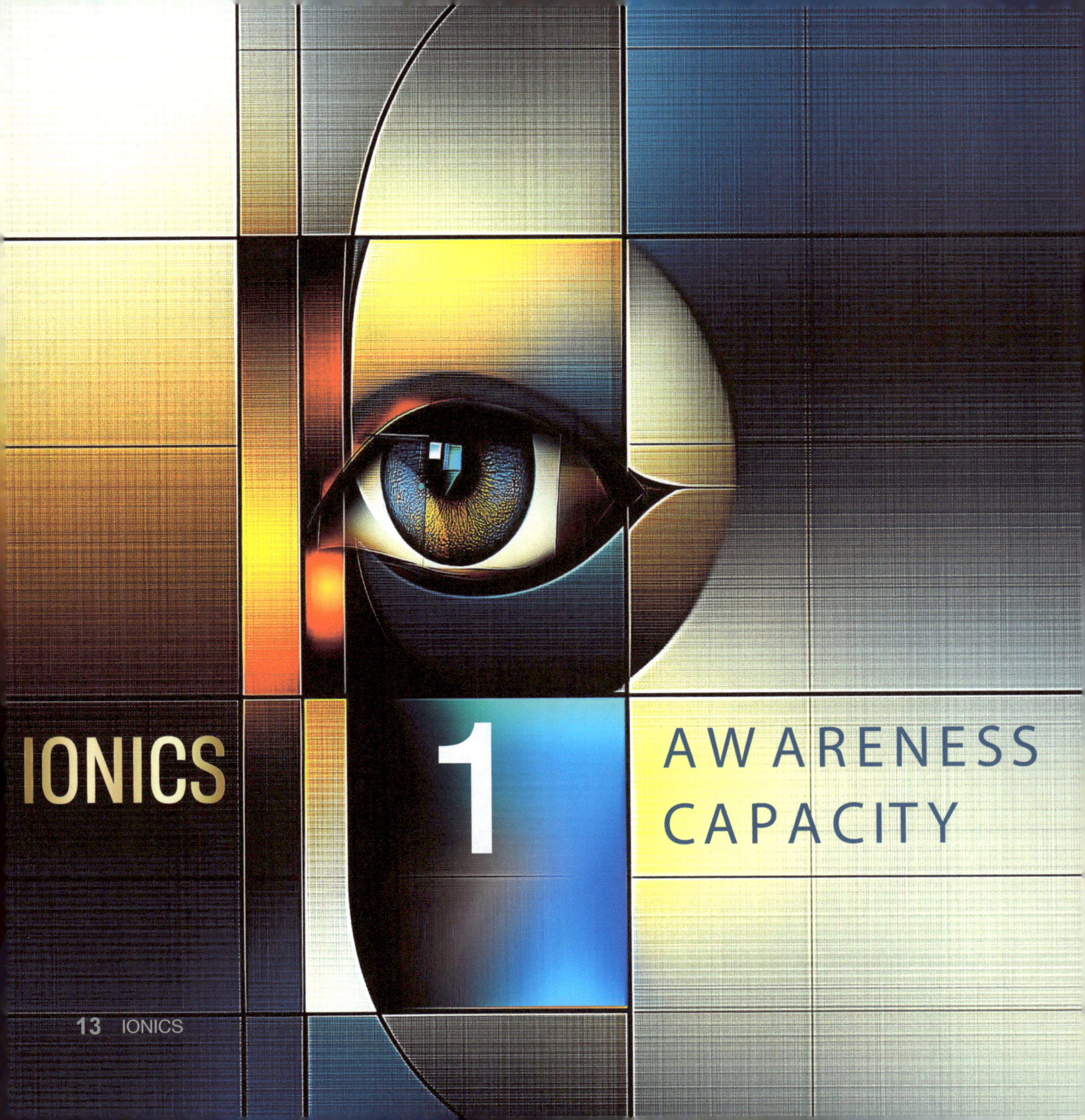

IONICS

1 AWARENESS CAPACITY

AWARENESS CAPACITY

Every person is inherently capable of higher states of being.

A peak experience might bring someone to an awareness level 5-10 times greater than their normal consciousness.

In reality, a human being's awareness potential is considerably greater

Awareness Metrics

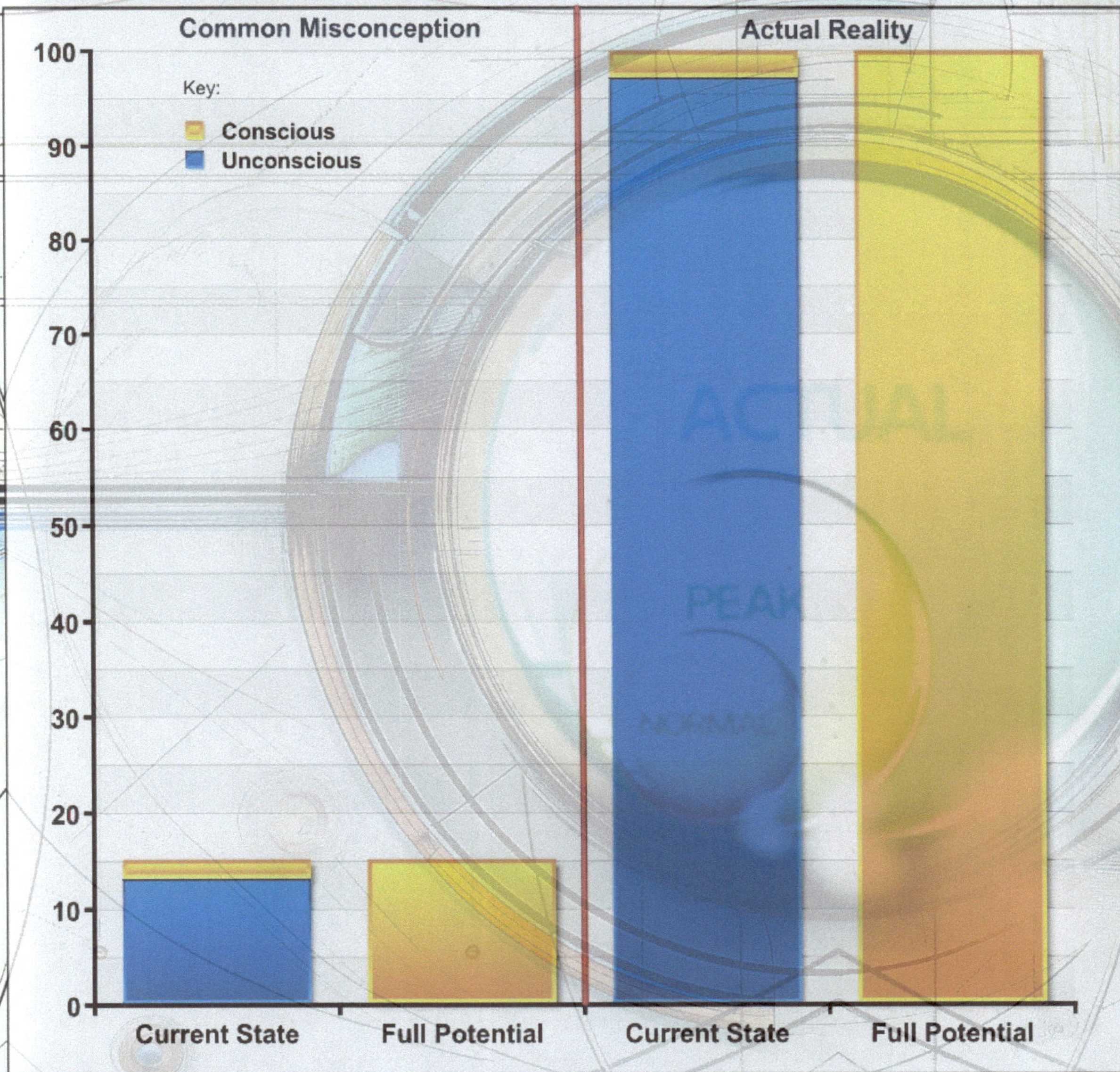

Our full potential for awareness is much larger than many people imagine.

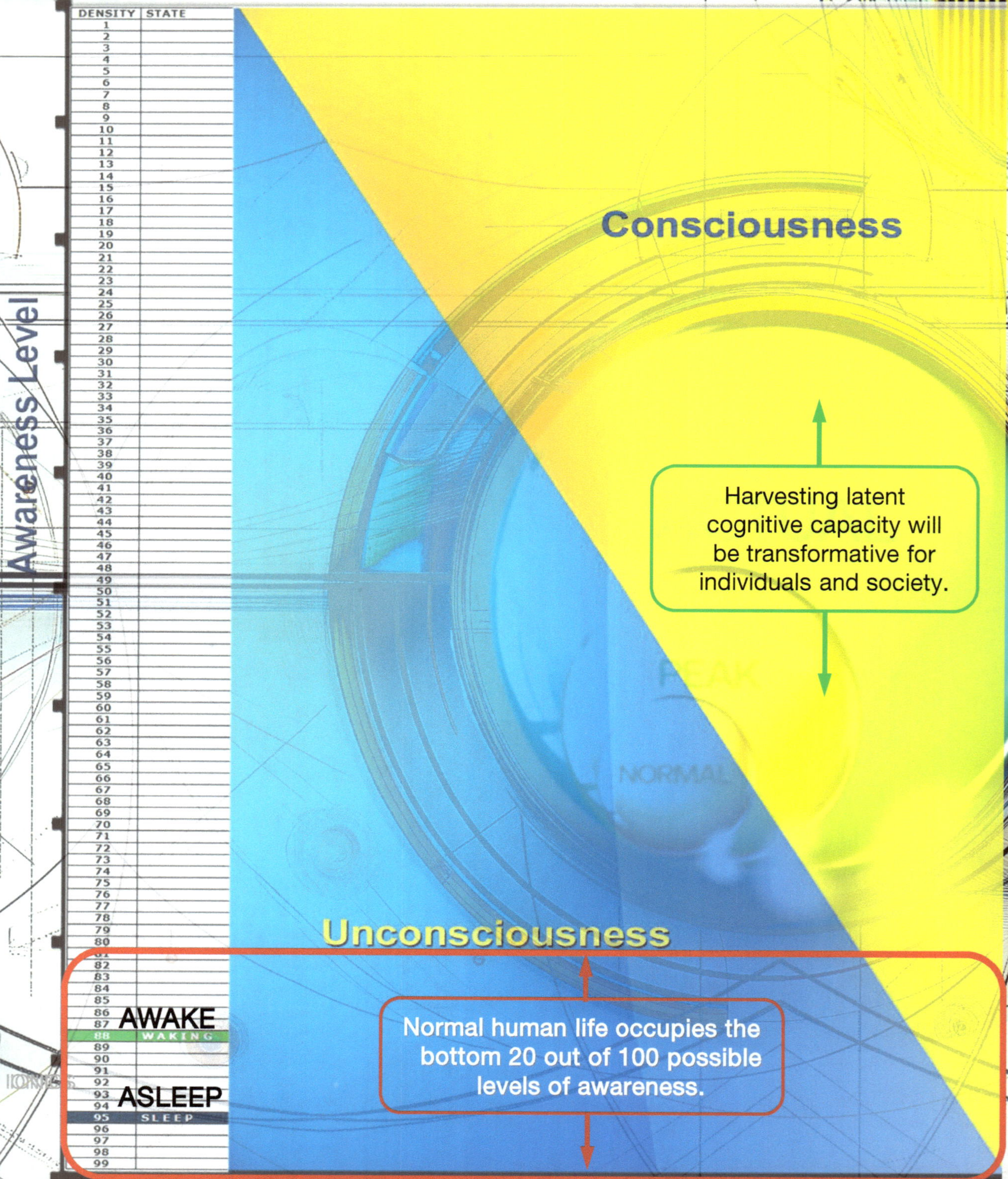
Awareness Level
DENSITY
STATE
1 2 3 4 5 6 7 8 9 10 11 12 13 14 15 16 17 18 19 20 21 22 23 24 25 26 27 28 29 30 31 32 33 34 35 36 37 38 39 40 41 42 43 44 45 46 47 48 49 50 51 52 53 54 55 56 57 58 59 60 61 62 63 64 65 66 67 68 69 70 71 72 73 74 75 76 77 78 79 80 81 82 83 84 85 86 87 88 89 90 91 92 93 94 95 96 97 98 99
Consciousness
Harvesting latent cognitive capacity will be transformative for individuals and society.
PEAK
NORMAL
Unconsciousness
AWAKE
WAKING
ASLEEP
SLEEP
Normal human life occupies the bottom 20 out of 100 possible levels of awareness.

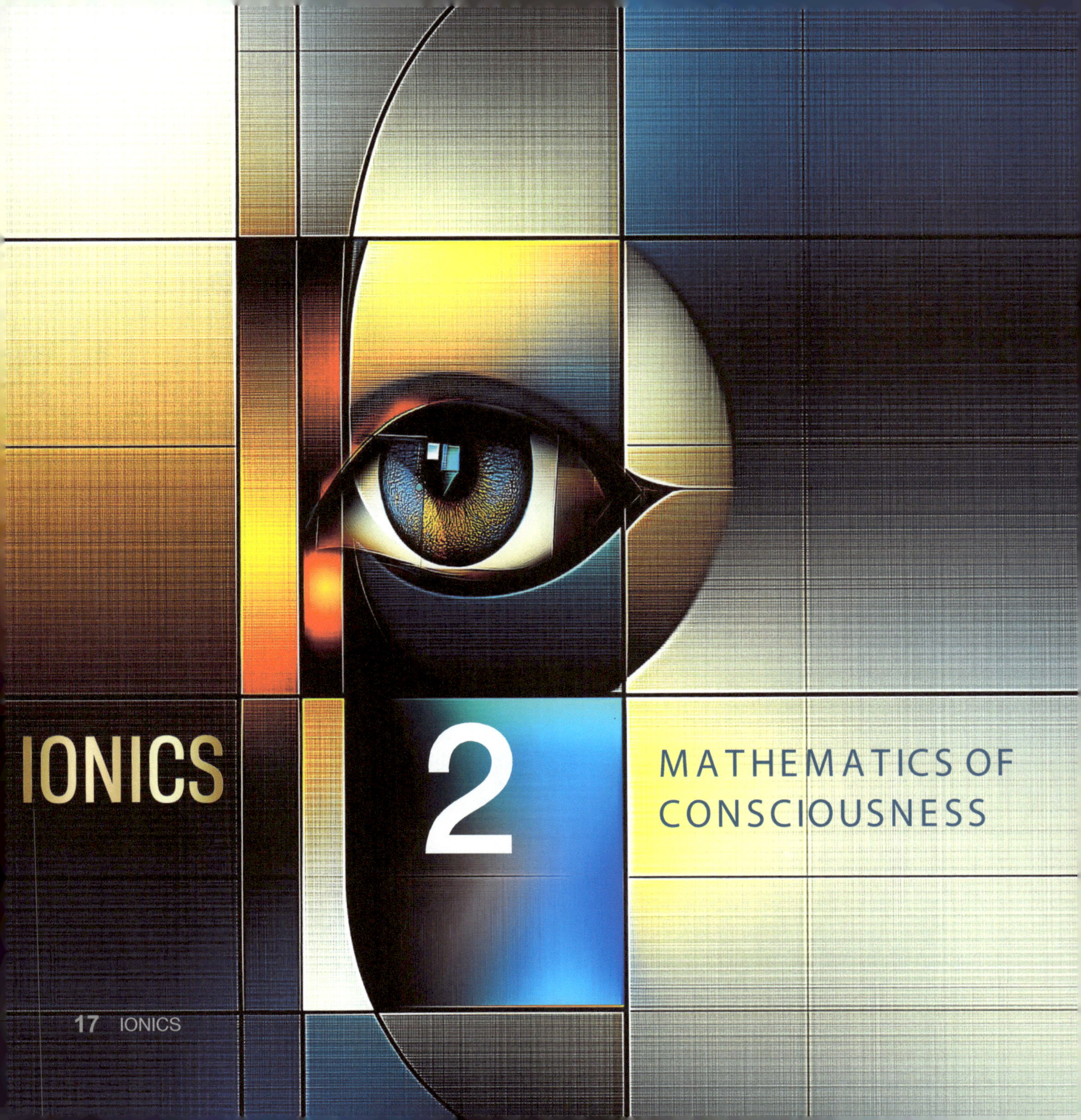

IONICS 2

MATHEMATICS OF CONSCIOUSNESS

AXIOMS and FORMULAS

The following axioms express the fundamental principles of awareness and its relationship with physical neurology in quantitative terms using scientific equations.

They reveal cardinal laws governing the dynamic interplay between awareness and the material universe.

The resulting body of knowledge enables scientific measurement and modulation of awareness states.

AXIOM 1

Awareness is a primary universal element.

Awareness is fundamental and irreducible.
The mind is not the brain.
Awareness envelopes the brain.

The axioms use the term "io" for awareness.

i representing conscious awareness
o signifying unconscious awareness.
Symbol: ϕ (phi)

AXIOM 2

Awareness consists of packets.

Io is composed of discrete packets called *iotons*, which are similar in principle to photons which comprise light.

Visualizing awareness in packets allows us to treat it numerically, opening the way to a mathematics of consciousness.

AXIOM 3

Iotons have two states: free and applied.

Free iotons make up conscious awareness. They are called *itons*.

Applied iotons comprise unconscious awareness. They are called *otons*.

An individual's awareness is a mixture of itons and otons.

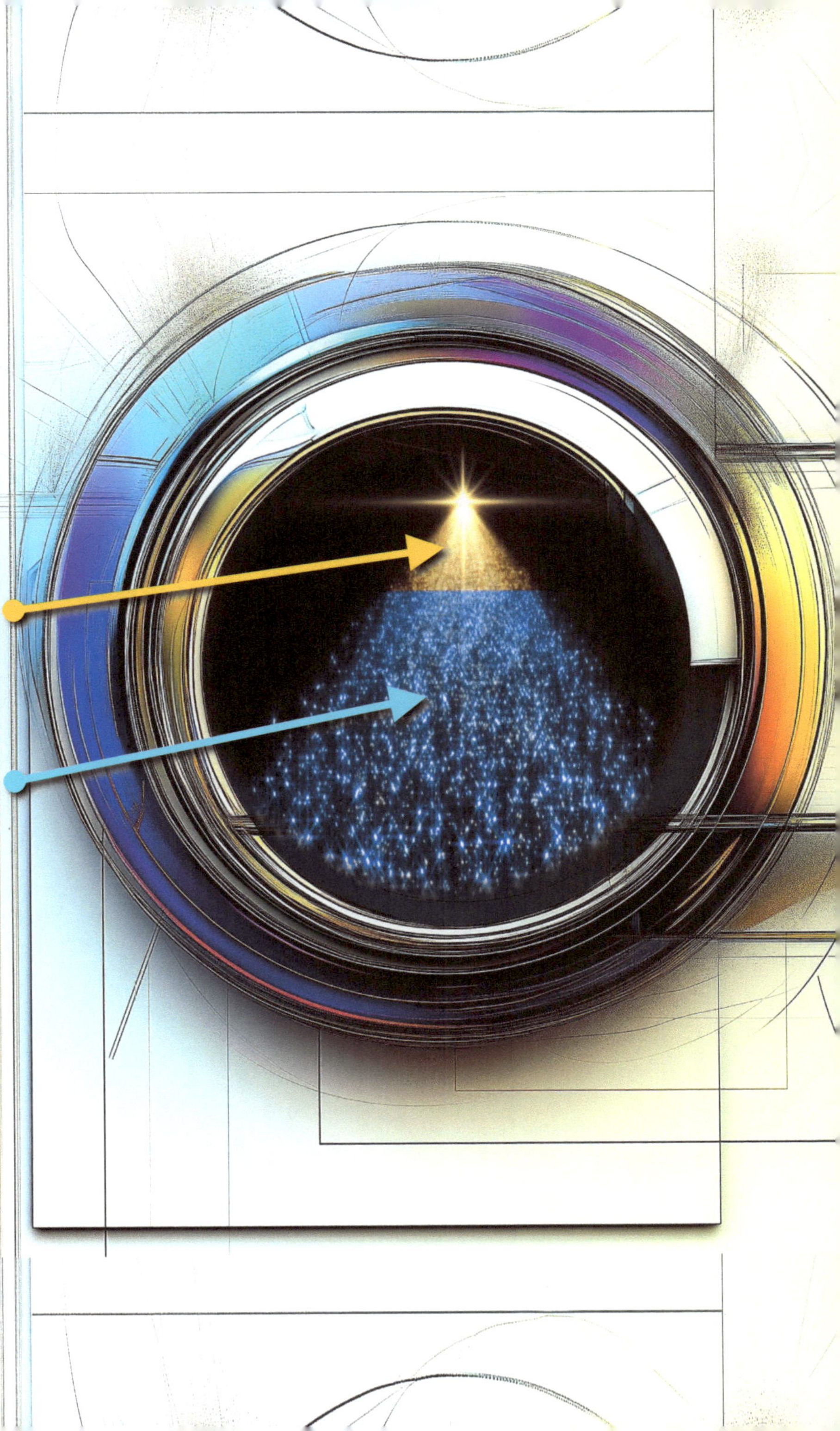

AXIOM 4

Free awareness is conscious and formless.

Free awareness, comprised of conscious itons, is changeless, motionless and stationary. The Vedic Upanishads describe this formless awareness as having no mass or energy, or location in space or time.

Corollary 1: Itons have no amplitude or wavelength.

Amplitude and frequency are forms which exist in space and time. Free awareness has a non-oscillating, static nature. This corollary has implications for enhancing human cognition.

AXIOM 5

Applied awareness is unconscious and has a form.

Applied awareness, composed of otons, is located in time and oscillates in time.

Itons change state to otons when they enter a form such as a memory. The applied otons permeate and animate the form, and in this process, become unconscious.

Otons power the functions of memory and physical life. When otons are removed from a memory, they revert back to their original free and conscious state as itons.

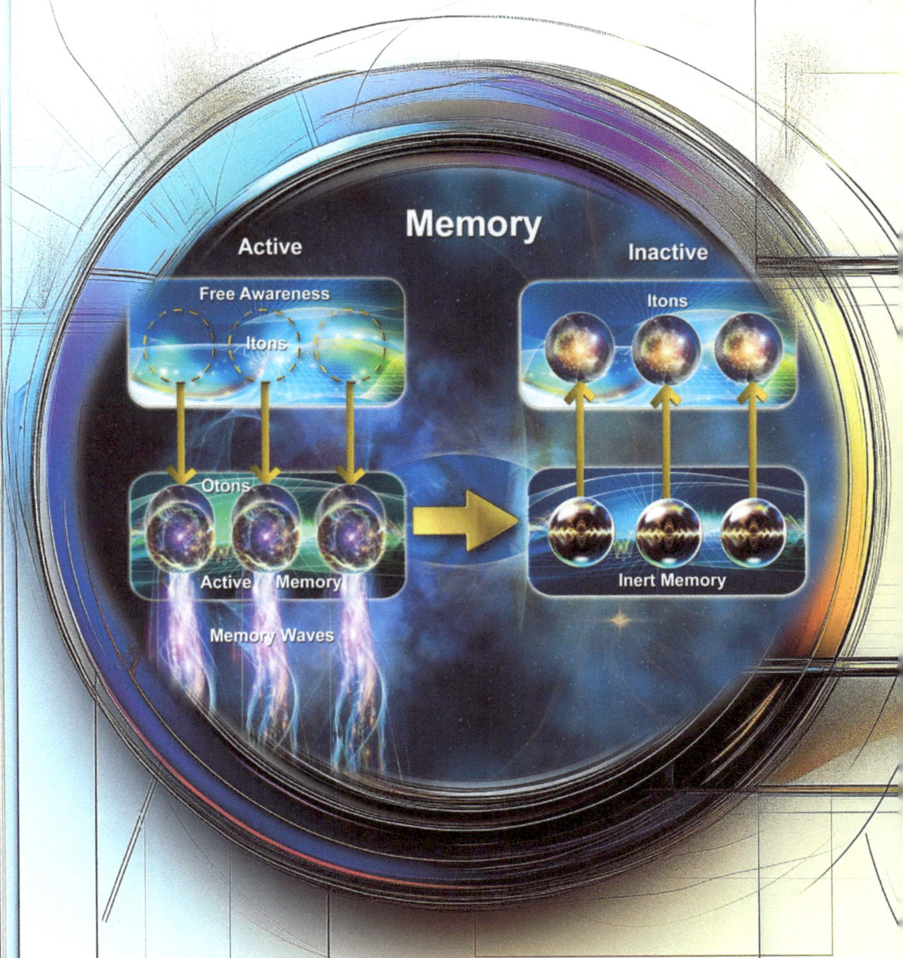

AXIOM 6

Law of Conservation of Awareness

Awareness cannot be created or destroyed; only changed from one form to another.

The volumes of consciousness and unconsciousness are reciprocals of each other; as one increases, the other decreases. However, an individual's total amount of awareness remains constant.

Formula 1: Law of Conservation of Awareness

$$I = (n - O)$$

where:

I = Conscious awareness itons
O = Unconscious awareness otons
n = Total iotons per individual

SUMMARY

Attribute	Iton	Oton
Io	Conscious	Unconscious
Type	Free	Applied
Time	Present	Past
Change	Motionless	Oscillates
Behavior	Particle	Wave
Function	Being	Knowing

IONICS

3 MIND/MATTER INTERFACE

MEMORY ACTIVATION

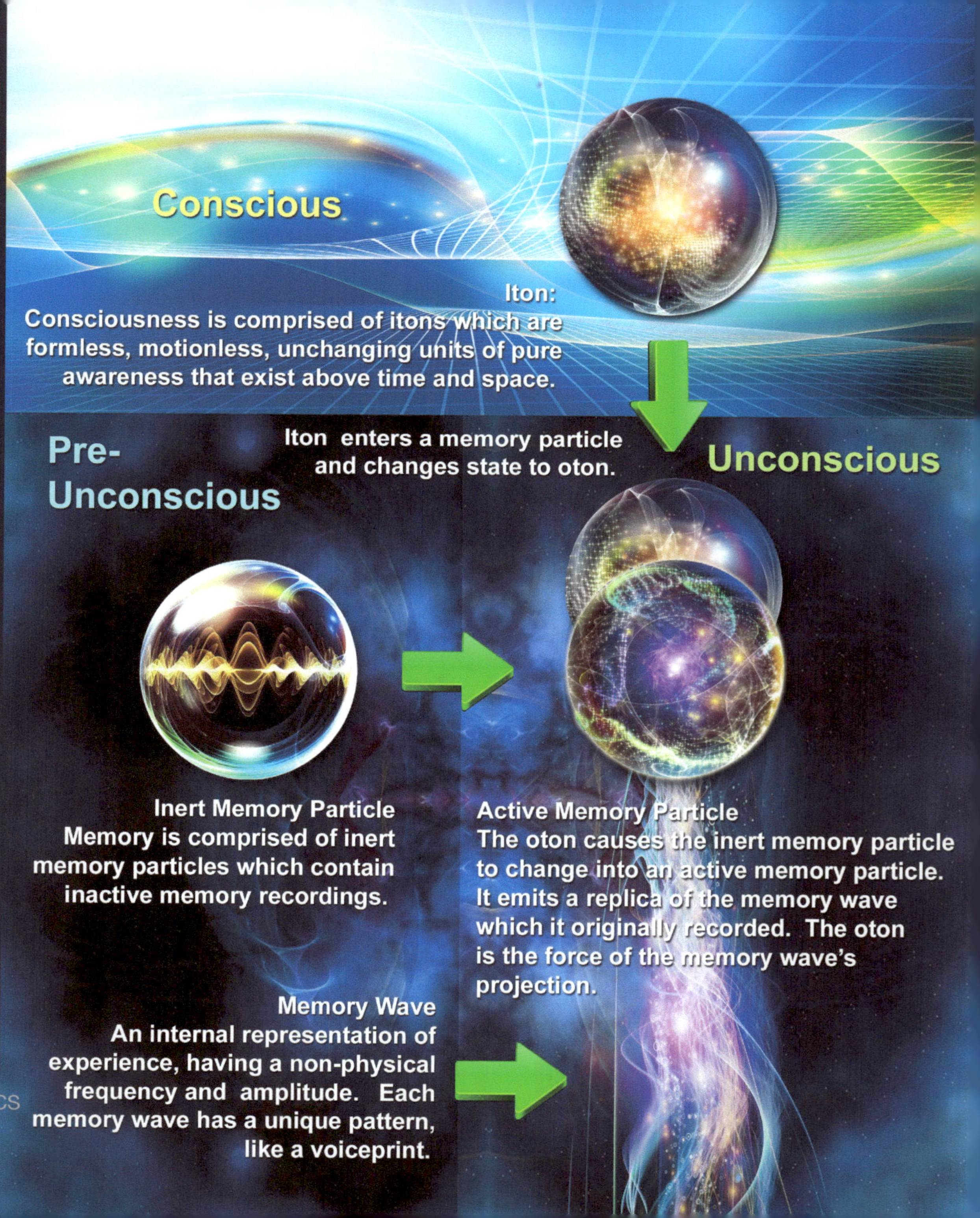

THE MIND/MATTER INTERFACE

Consciousness condenses through 5 layers to form the living universe.

Active memory particles (top) emit *memory waves* (gold) which are propagated through *space* (blue).

Space waves in turn generate identical patterns in cellular-level brainwave voltage fields marked *neurowaves* (green).

Neurowave voltage propels *electrons* through *neurons*, powering the physical functions of life.

This top-down causality describes behavior. The order is reversed for perception.

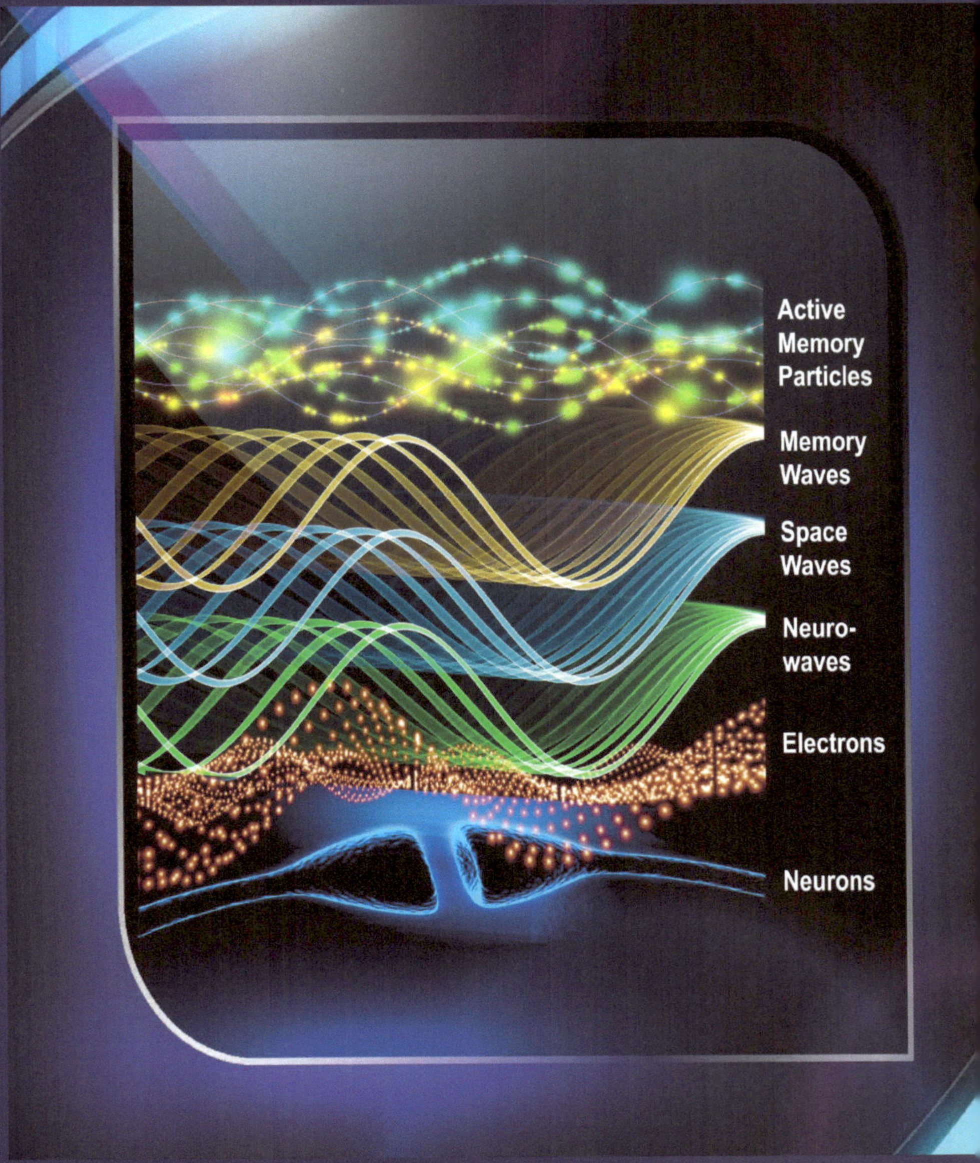

AXIOM 7

The Mind / Matter Interface

The interface between awareness and neurology comprises five layers:

1. Free awareness (being)
2. Information (knowing)
3. Space zero-point fields
4. Neurowave fields
5. Neural network activity

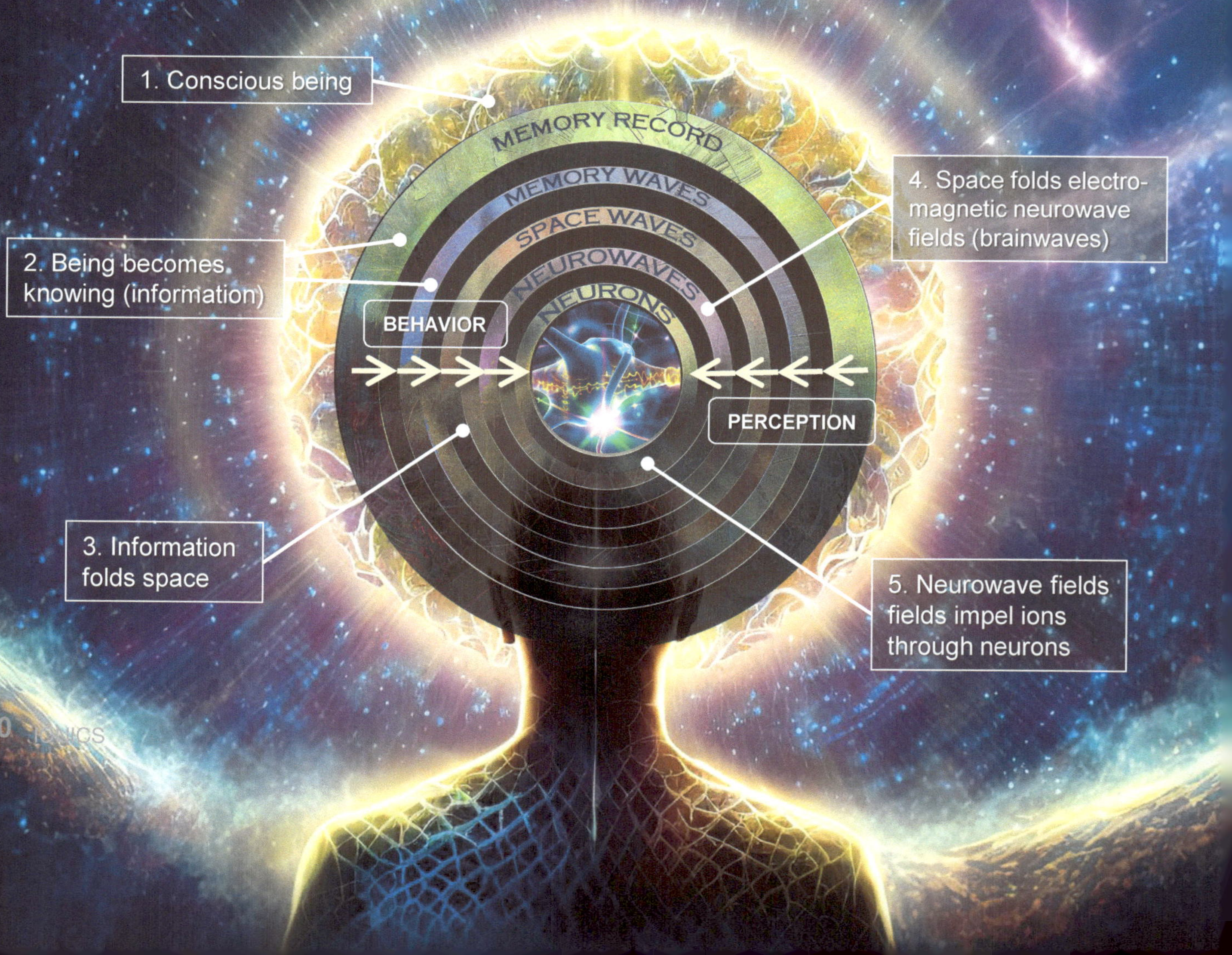

Brainwaves are the physical manifestation of active unconscious memories.

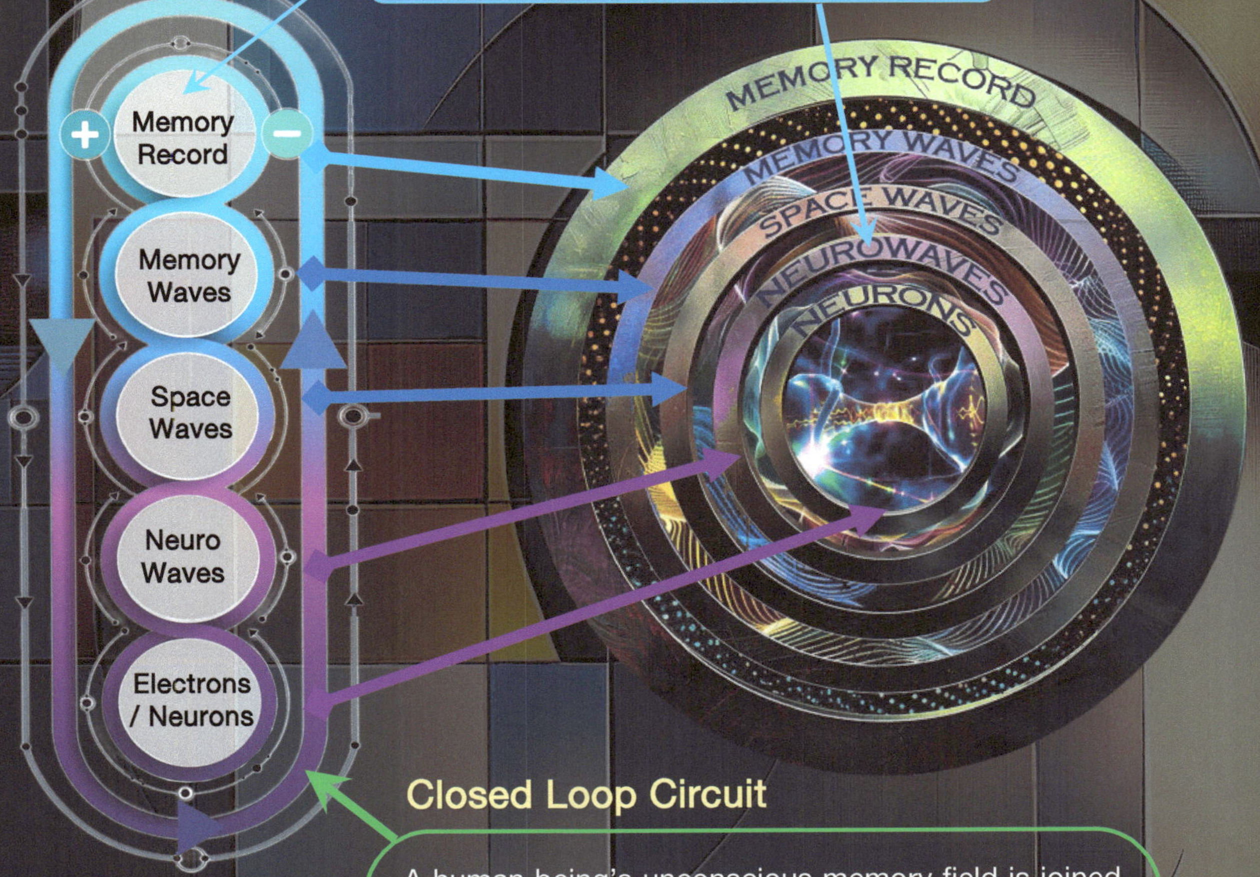

Closed Loop Circuit

A human being's unconscious memory field is joined in an electrical circuit with their body's neurology.

This closed loop circuit culminates in neurowave fields as the fundamental interface zone between awareness and physical neurology.

Communication across this interface must follow the laws which govern all electromagnetic fields.

MIND/MATTER INTERFACE DYNAMICS

Layer 1 - Memory Record

A memory record is a snapshot of memory waves arising from neurowaves in a given moment. Each recorded perception is encoded by a memory particle, which emits a memory wave when activated. (1)

The memory record is a blueprint for re-generating the memory waves it originally recorded, like a film frame. It is inert until activated by otons. Activation reproduces the neurowave snapshot, which is overlayed on current neurology.

(1) e.g., visual, auditory, spatial orientation, temperature, sensation, taste, smell, tactile, emotion

Memory Record Charge

A memory record's charge is proportional to the total number of otons enclosed by the record.

Memories can be fully or partially reactivated. Only active memories have measurable charge. Inert memories have no charge, only potential.

The number of otons contained in the record varies with the amplitude of the original memory wave signatures it recorded.

Layer 2 – Memory Waves

Memory waves are vector fields with frequency, amplitude and a direction to the neural circuits they arose from.

The strength of the memory wave field is proportional to the enclosing memory record charge.

A memory wave's amplitude varies with the number of otons animating the memory record emitting it.

Layer 3 – Space Waves

- The strength of the space wave field is proportional to the enclosing memory wave field charge.

Memory waves induce oscillations in the fabric of space which, in turn, affect the brain's electromagnetic fields at a cellular level. (1)

(1) The best language we currently have to describe these spatial vibrations is "quantum fluctuations in the zero-point energy fields of space." Today's terms are placeholders for phenomena we know exist but do not fully understand. As science progresses, better nomenclature will emerge.

Layer 4 – Neurowaves

The strength of the neurowave field is proportional to the enclosing space field charge.

Layer 5 – Neurons

The **electrical current** passing through a neural network is proportional to the enclosing neurowave field charge.

MIND/MATTER INTERFACE DYNAMICS

Summary - Amplitude

Layer 1 – Memory Record: The memory record charge is proportional to the total number of otons enclosed by the record.

Layer 2 – Memory Waves: The strength of the memory wave field surrounding a space field is proportional to the enclosing memory record charge.

Layer 3 – Space: The strength of the space wave field surrounding a neurowave field is proportional to the enclosing memory wave field charge.

Layer 4 – Neurowaves: The strength of the neurowave field surrounding a neural network is proportional to the enclosing space wave field charge.

Layer 5 – Neural Network: The strength of the electric field passing through a neural network is proportional to the enclosing neurowave field charge.

IONICS

4 COGNITIVE ENHANCEMENT

ENHANCING AWARENESS

Otons constitute a large reservoir of encapsulated awareness which can be tapped for cognitive enhancement.

Itons can be recaptured by shifting otons out of animating unconscious information.

In this way, dormant awareness can be restored back to consciousness by retrieving it from inessential memories.

Disrupting Memory

An individual can be disengaged from animating unconscious information by altering the memory wave frequency signatures which carry the information.

Disrupting a memory wave's integrity causes it to lose its information content and discharge the associated otons.

Co-Residence

Each memory wave is co-resident with a corresponding space wave and neurowave component in a closed loop circuit.

Changing either of these components will reshape the memory wave.

Neurowave Disruption

Neurowaves arising from active unconscious memories overlay themselves onto the same locations in the brain from which they were originally recorded.

Changing neuron timing in any of these locations will disrupt the neurowave's integrity, collapsing the associated memory wave and releasing otons.

Neuroanatomy Targets

The Dysaffective Connectome is a network of neural pathways controlling the storage of dysfunctional emotional memories, the processing of negative emotional experiences, and the generation of maladaptive emotional responses.

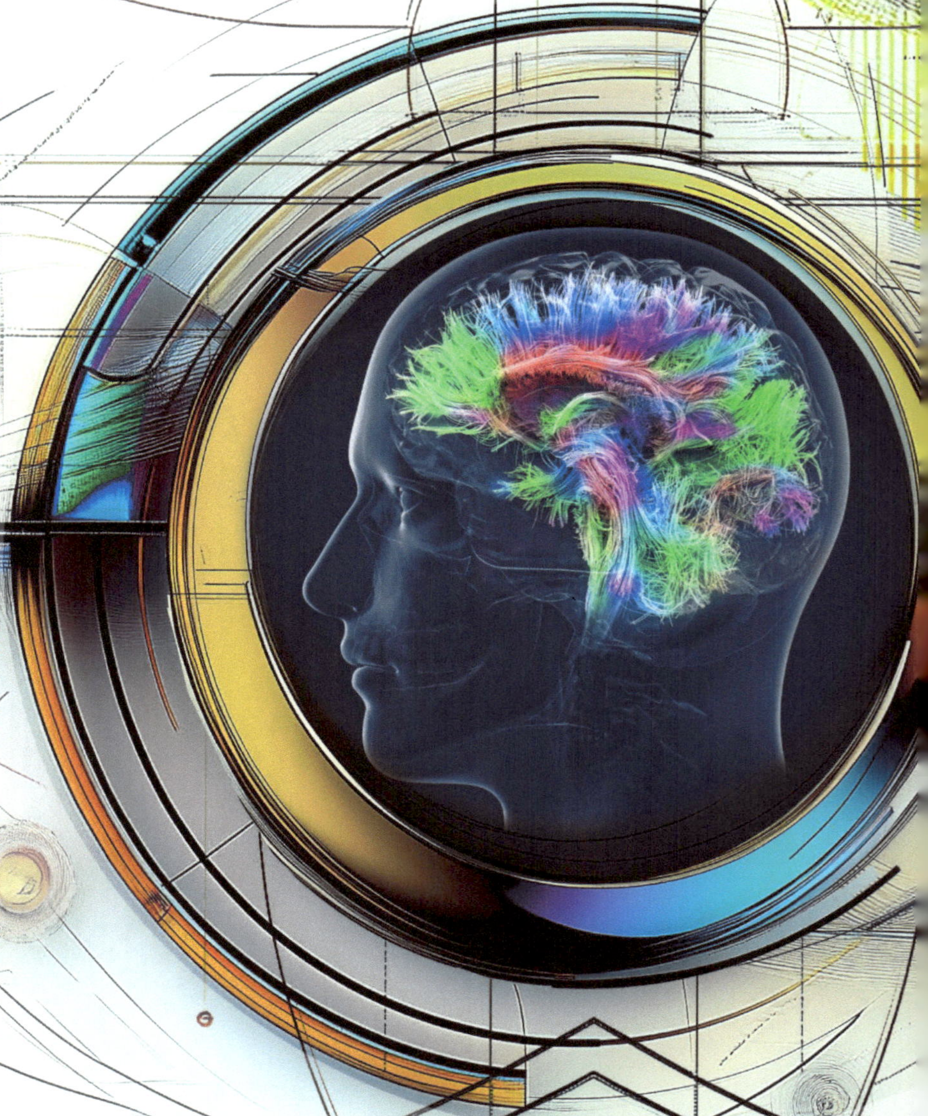

Its hubs are located in the amygdala, hippocampus, ventral posterior cingulate cortex, and, to a lesser extent, the insula and thalamus.

Lowering cortical activity in these brain regions has reduced anxiety and improved attention in animals. In humans, attenuating neural signaling in this network can restore awareness encapsulated in aversive memories.

IONICS

5 NEURO-ENGINEERING

Target

Neurowaves are the most accessible component in the closed loop stack to modify. They can be altered by changing the electrical properties of the neurons which carry them through the brain.

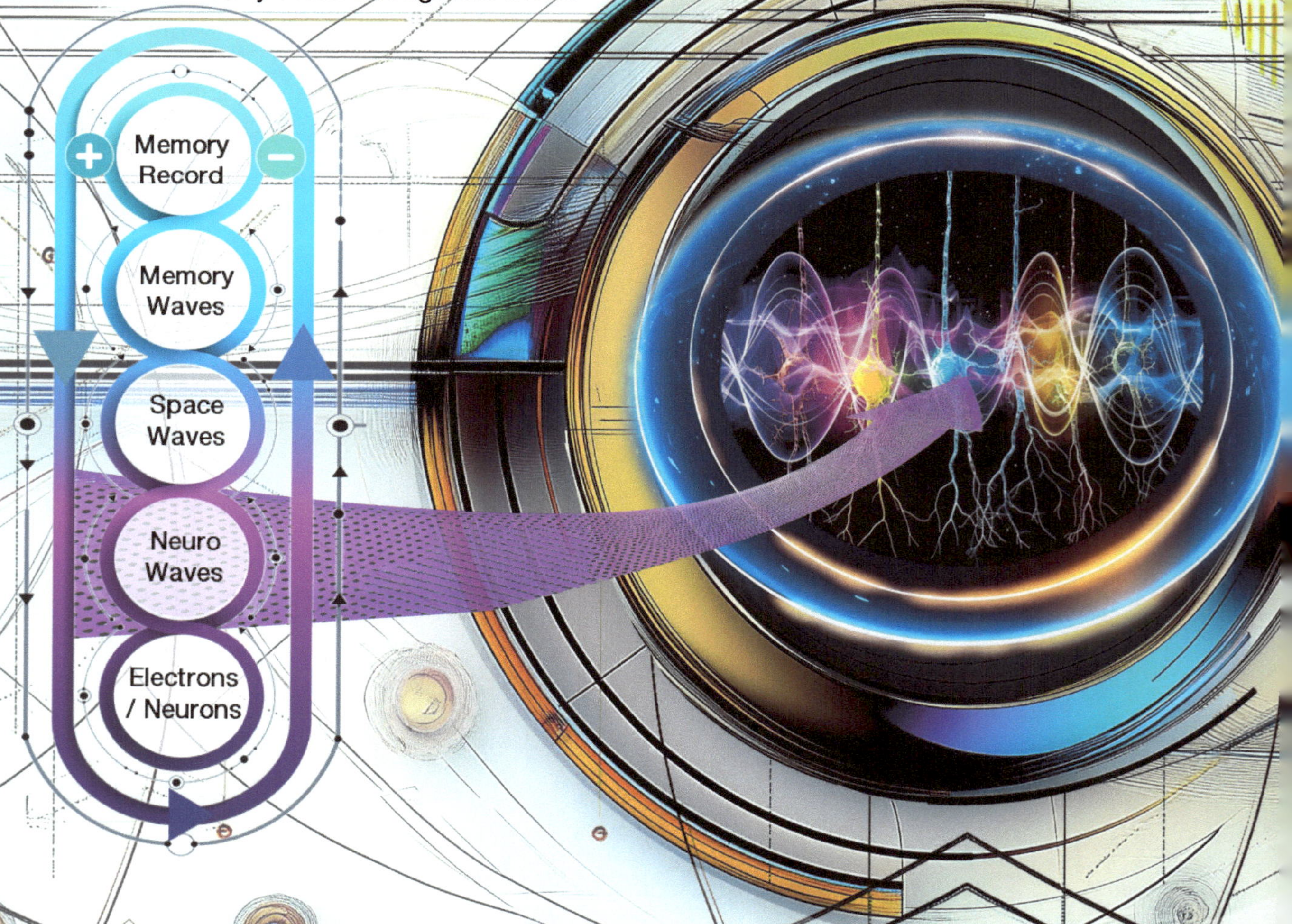

Neurowave Modification

Neurowaves travel through chains of neurons in the brain.

These neurons all have set timings which govern how fast the waves pass through them.

Changing the timing of neurons in the chain disrupts the overall frequency signature of the neurowave traveling through it.

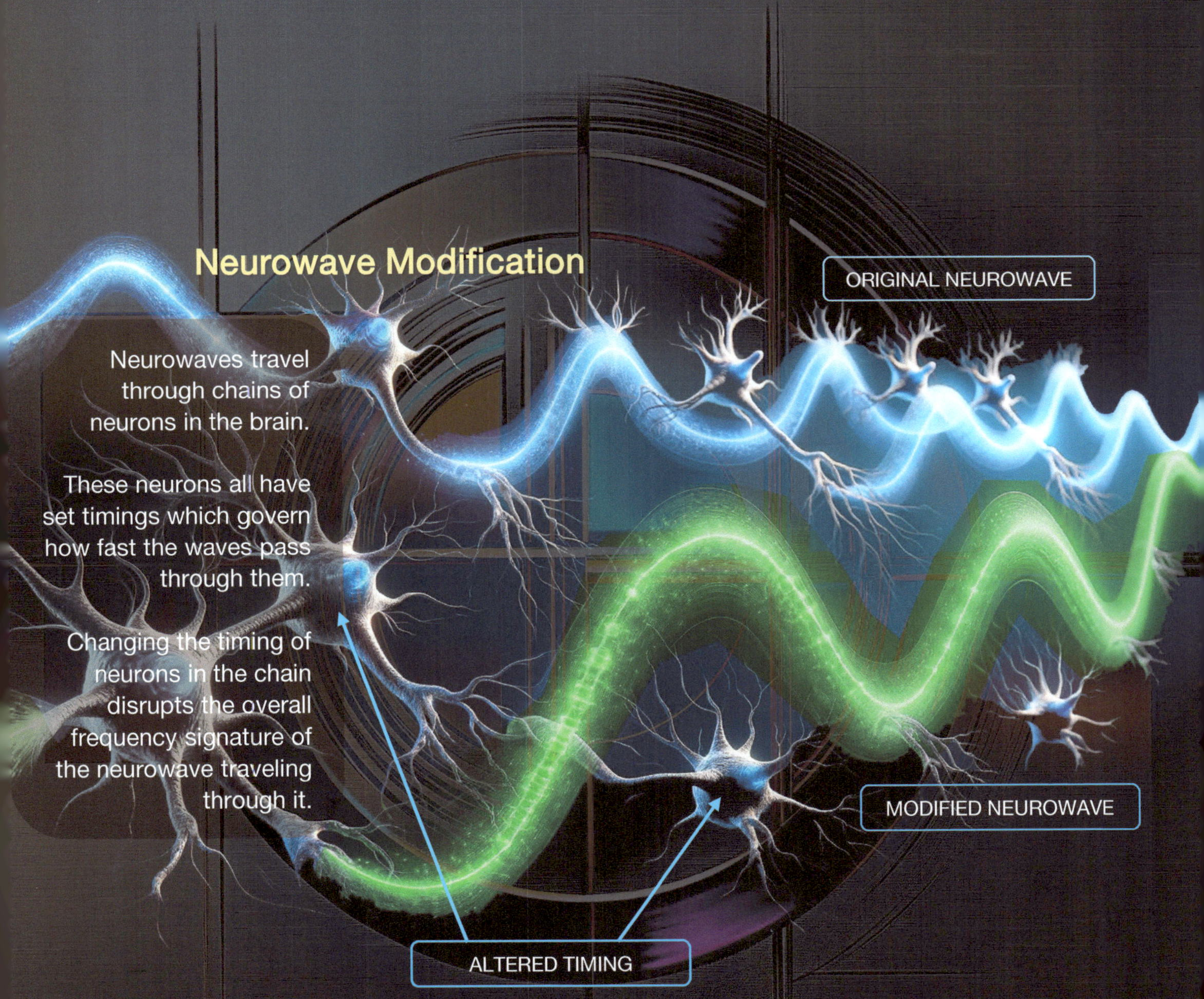

Neuron Timing

When a neuron receives a sufficient number of incoming activation signals (neurotransmitters) it fires an electrical pulse to the next neuron.

How quickly the neuron fills up determines how rapidly it sends an action potential (or "spike") onto the next neuron.

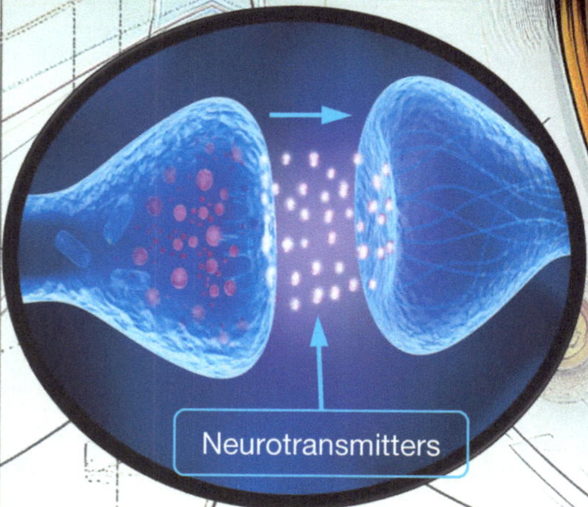

Method of Action

A neuron acts electrically like a capacitor. (1) It absorbs incoming charge at a rate determined by signals that it receives through certain excitatory neurotransmitter receptors.(2)

A neuron's receptivity to neurotransmitters can be genetically modified, thereby changing its activation time and firing rate.

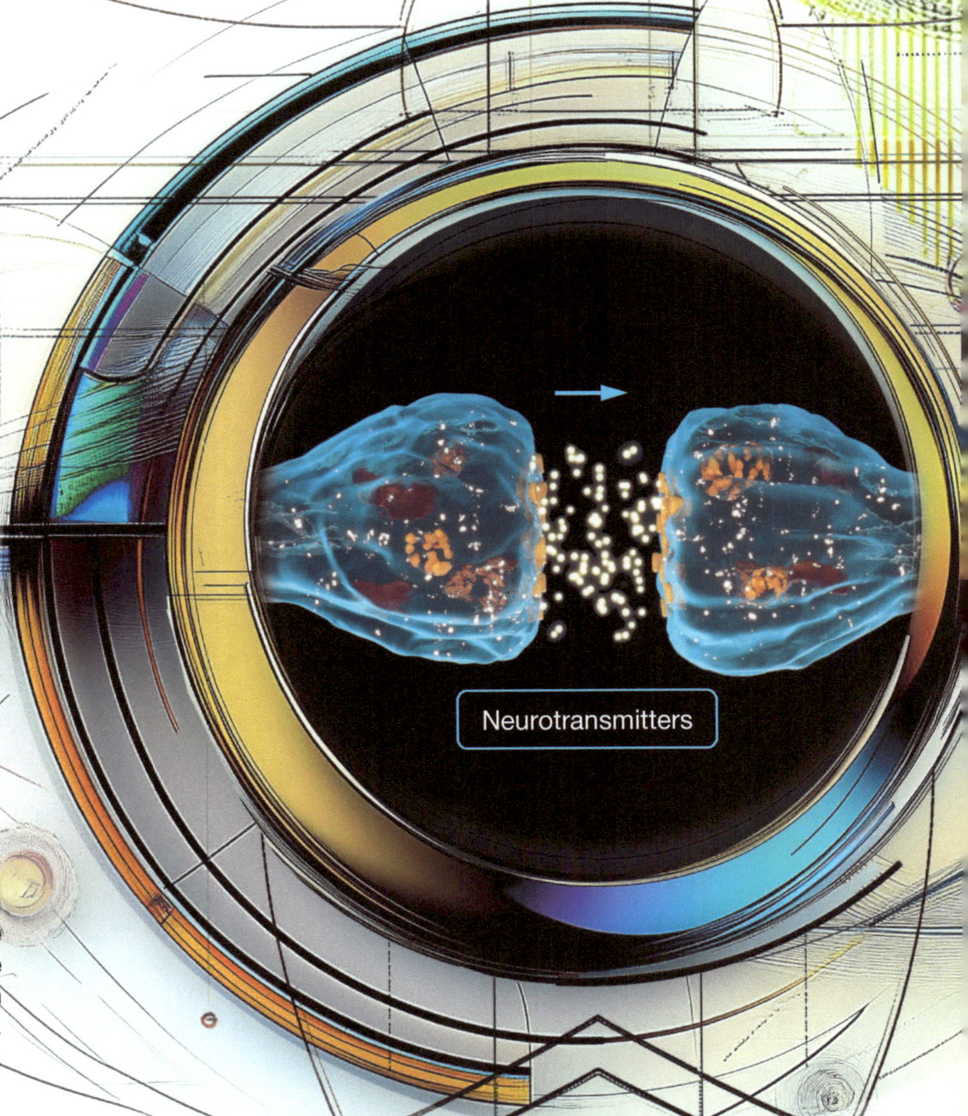

(1) A capacitor receives and stores incoming electrons until it reaches its storage threshold. It then discharges the electrons in a burst to the next element in the circuit.

(2) e.g., serotonin and dopamine.

Neuron Electrodynamics

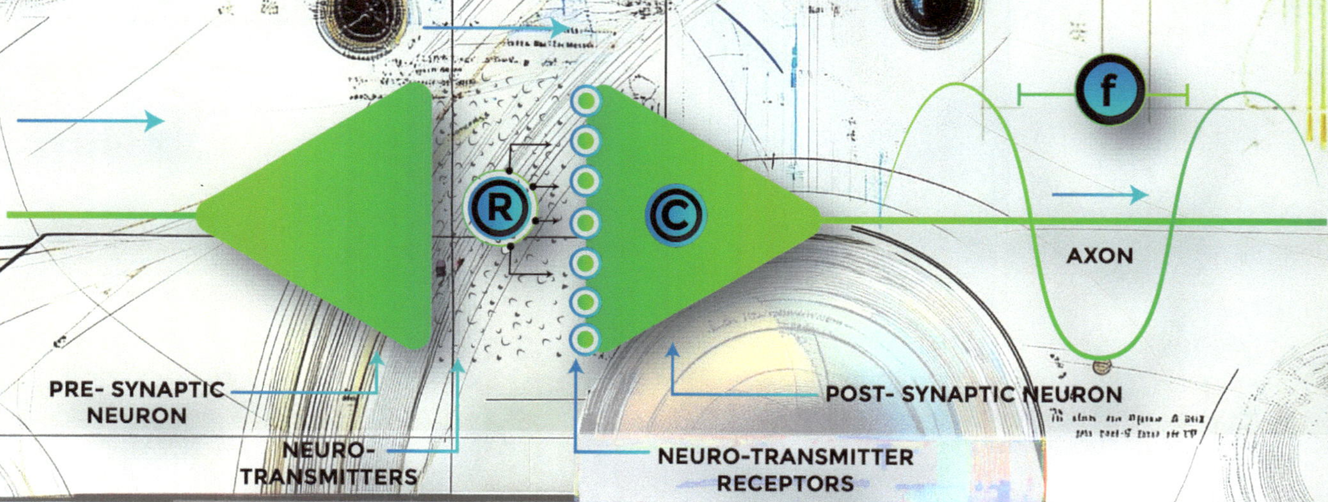

If these neurons were in an electrical circuit, we would write the formula:

$$f = \frac{1}{2\pi RC}$$

Resistance **R** varies with the number of receptors receiving activation signals.
Capacitance **C** is the neuron's ability for storing electrical charge.
Frequency **f** is the number of spikes per second or the wavelength between peaks.

Capacitance is basically a constant. Thus, resistance R determines frequency f.

Resistance R can be precisely controlled through genetic or transcriptional engineering to change the number of open receptor channels for receiving neurotransmitter signals, thereby controlling frequency of the output.

Note: This equation applies to excitatory receptors, i.e., certain serotonin, glutamate and dopamine subtypes.

IONICS

6 FORMULAS

IONICS FORMULAS

1. Memory record

1. charge
2. polarity

2. Wave fields

a) Frequency
- a) Behavior
  - a) Memory waves
  - b) Space waves
  - c) Neurowaves
- b) Perception
  - a) Memory waves
  - b) Space waves
  - c) Neurowaves

b) Amplitude
- a) Memory waves
- b) Space waves
- c) Neurowaves

3. Neural network
- a) Voltage

A. Memory Record Charge

The memory record charge is proportional to the total number of otons enclosed by the record.

$$\oint \mathbf{MR}\cdot\text{-}d\mathbf{a} = Qenc \,/\, \epsilon 0$$

Where:

$\oint$ denotes the closed surface integral (1)

MR is the memory record charge

da is the infinitesimal area vector on the closed surface, pointing outward (2)

$\boldsymbol{Qenc_O}$ is the number of otons enclosed by the memory record

$\boldsymbol{\epsilon 0_{MR}}$ is a permittivity constant defining how otons enter a memory record

(1) A closed surface integral ($\oint$) is a calculus tool used to sum up a quantity over a closed surface.

(2) A differential area element (***da***) is a calculus tool used to break down complex surfaces into manageable, infinitesimally small pieces for integration purposes.

1B. Memory Record Polarity

Memories are recordings of oscillating neurowave signatures in neurology. Since these waves have polarity, their recordings also have polarity.

The memory's polarity drives energy through a closed loop circuit from the plus to the minus side of the memory as shown. The strength of this polarity (plus/minus difference) is the force of its projection through the circuit.

The memory record's polarity or charge is proportional to the amplitude of its recordings. The force or voltage of the closed loop circuit varies with the amplitude of the memory's recorded waves.

All memory records have polarity. There are no memory monopoles.

The memory flux through any closed loop neural circuit is zero.

$$\nabla \cdot \mathbf{M} = 0$$

Where:
∇ is the divergence operator
M is the memory field

In other words, no waves diverge from or flow out of the closed loop circuit. It is self-contained.

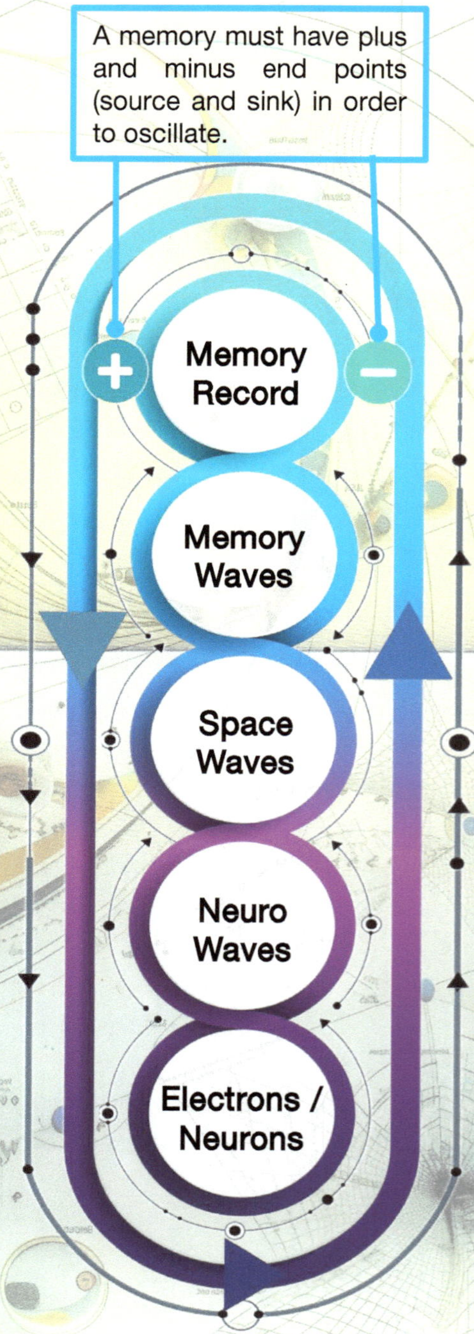

2. Wave Fields

The mind/matter interface's three wave fields are illustrated here three dimensionally.

Behavior waves emanate from the memory record and terminate in neurology.

Perception waves originate in neurology and terminate in the memory record.

There is an instantaneous, continuous feedback loop between behavior and perception.

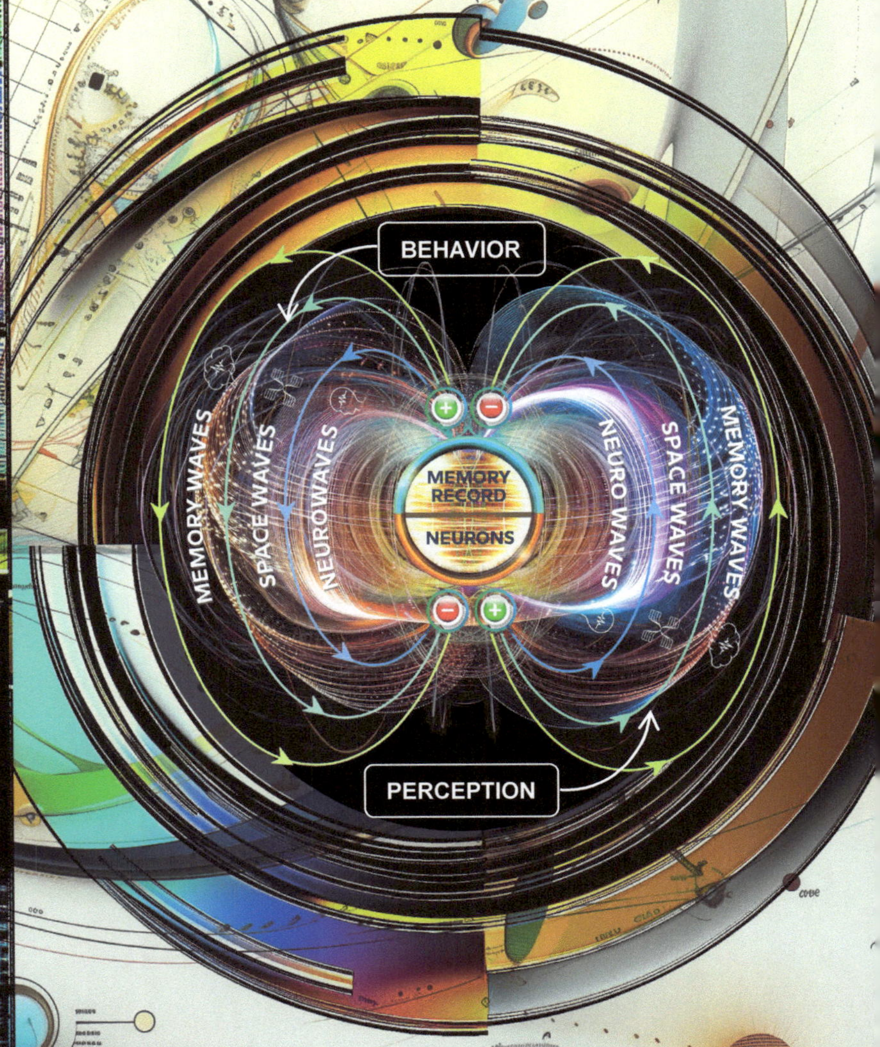

Wave Functions Cross Section

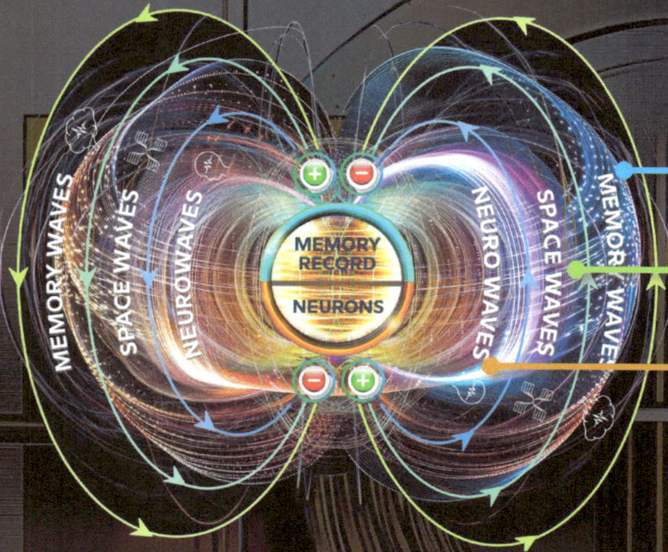

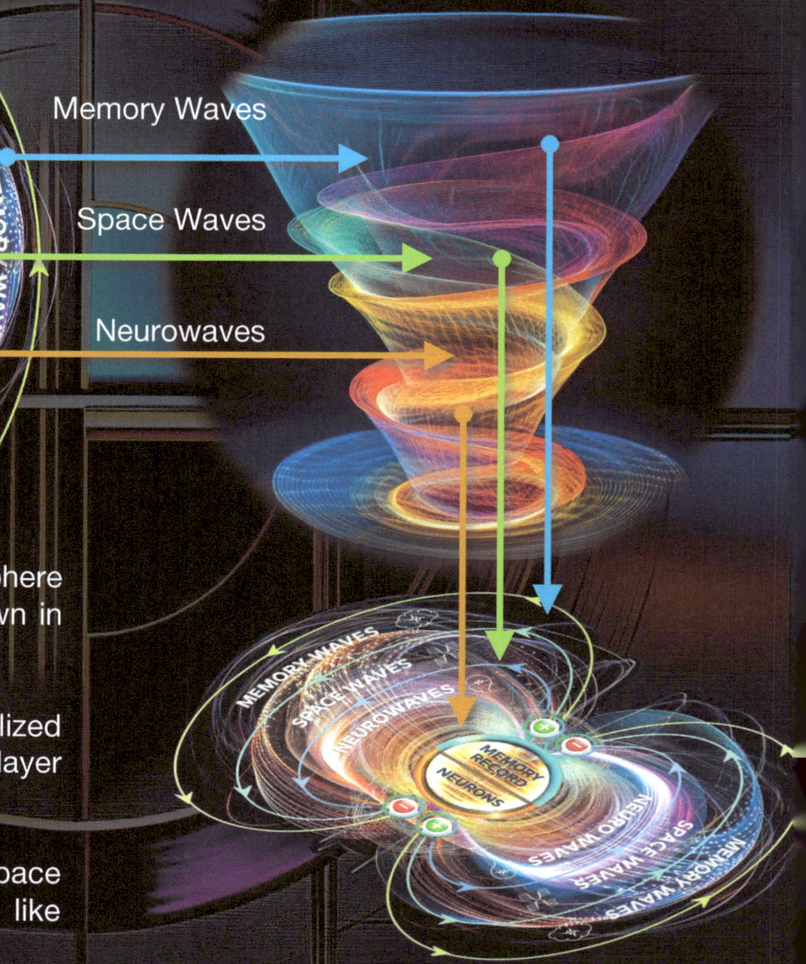

The three types of waves form a sphere around the memory record, as shown in the cross section here.

This architecture can also be visualized as a funnel with each successive layer having greater density.

Memory waves push against the space membrane surrounding a being like sound waves vibrate an eardrum.

Wave Field Formulas - Frequency

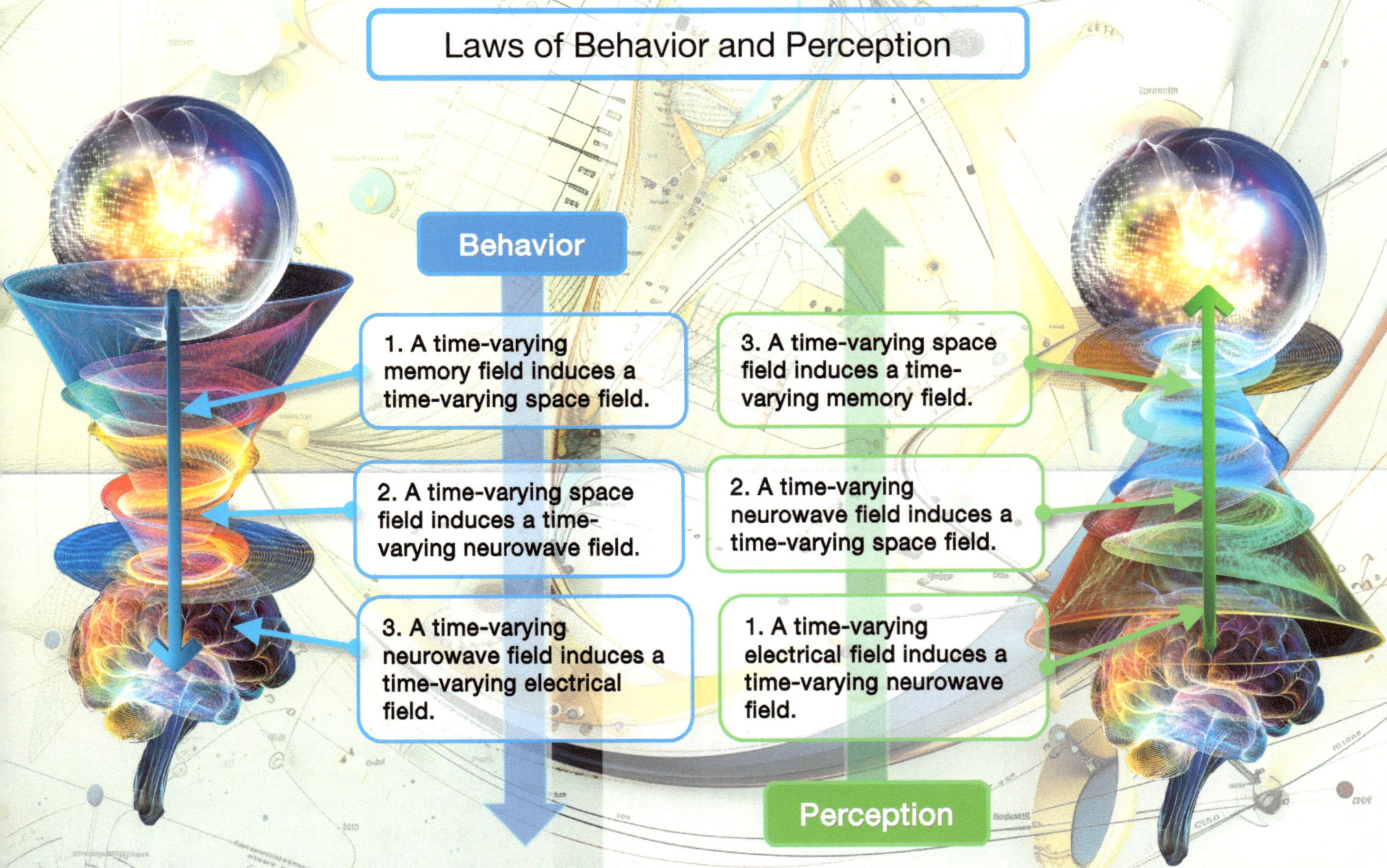

First Behavior Equation

Behavior

1. A time-varying memory field induces a time-varying space field.

$$\frac{\partial \boldsymbol{S}}{\partial t} = k\frac{\partial \boldsymbol{M}}{\partial t}$$

Where:
M: memory wave
S: space wave
t: time
∂: partial derivative
k: proportionality constant

The time variation of space field S is directly proportional to the time variation of memory field M.

- $\partial S/\partial t$ is the time derivative of field S, representing how S changes over time.
- $\partial M/\partial t$ is the time derivative of field M, representing how M changes over time.
- k is a proportionality constant that relates the rate of change of S to the rate of change of M.
- ∂ indicates a partial derivative, meaning the derivative is taken with respect to one variable (time t) while holding other variables constant.

Laws of Behavior

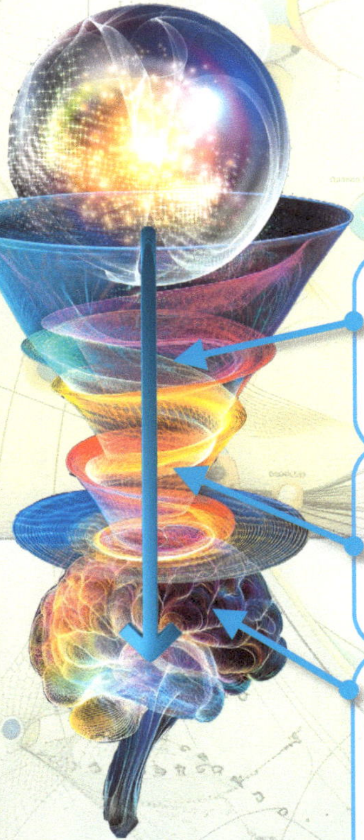

Behavior

1. A time-varying memory field induces a time-varying space field.

$$\frac{\partial S}{\partial t} = k\frac{\partial M}{\partial t}$$

2. A time-varying space field induces a time-varying neurowave field.

$$\frac{\partial N}{\partial t} = k\frac{\partial S}{\partial t}$$

3. A time-varying neurowave field induces a time-varying electrical field.

$$\frac{\partial E}{\partial t} = k\frac{\partial N}{\partial t}$$

Where:
M: memory wave
S: space wave
N: neurowave
E: electrical field
t: time
∂: partial derivative
k: proportionality constant

- $\partial S/\partial t$ is the time derivative of field S, representing how S changes over time.
- $\partial M/\partial t$ is the time derivative of field M, representing how M changes over time.
- $\partial N/\partial t$ is the time derivative of field N, representing how N changes over time.
- $\partial E/\partial t$ is the time derivative of field E, representing how E changes over time.

- k is a proportionality constant that relates the rate of change of the first field to the rate of change of the second field.
- ∂ indicates a partial derivative, meaning the derivative is taken with respect to one variable (time-t) while holding other variables constant.

Laws of Perception

Where:
M: memory wave
S: space wave
N: neurowave
E: electrical field
t: time
∂: partial derivative
k: proportionality constant

- $\partial S/\partial t$ is the time derivative of field S, representing how S changes over time.
- $\partial M/\partial t$ is the time derivative of field M, representing how M changes over time.
- $\partial N/\partial t$ is the time derivative of field N, representing how N changes over time.
- $\partial E/\partial t$ is the time derivative of field E, representing how E changes over time.

$$\frac{\partial \boldsymbol{M}}{\partial t} = k\frac{\partial \boldsymbol{S}}{\partial t}$$

3. A time-varying space field induces a time-varying memory field.

$$\frac{\partial \boldsymbol{S}}{\partial t} = k\frac{\partial \boldsymbol{N}}{\partial t}$$

2. A time-varying neurowave field induces a time-varying space field.

$$\frac{\partial N}{\partial t} = k\frac{\partial \boldsymbol{E}}{\partial t}$$

1. A time-varying electrical field induces a time-varying neurowave field.

Perception

Wave Field Formulas - Amplitude

Layer 1 – Memory Field: The strength of the memory wave field is proportional to the enclosing memory record charge.

$$\oint \mathbf{M}\cdot\text{-}d\mathbf{a}= Qenc_{MR} / \epsilon 0_{M}$$

Layer 2 – Space Field: The strength of the zero point space field is proportional to the enclosing memory field charge.

$$\oint \mathbf{S}\cdot\text{-}d\mathbf{a}= Qenc_{M} / \epsilon 0_{S}$$

Layer 3 – Neurowave Field: The strength of the neurowave field is proportional to the enclosing space field charge.

$$\oint \mathbf{N}\cdot\text{-}d\mathbf{a}= Qenc_{S} / \epsilon 0_{N}$$

Where:
$\oint$ denotes the closed surface integral
M is the memory field amplitude
S is the space field amplitude
N is the neurowave field amplitude
da is the infinitesimal area vector on the closed surface, pointing outward
$Qenc_{MR}$ is the total charge enclosed by the memory record
$Qenc_{M}$ is the total charge enclosed by the memory wave field
$Qenc_{S}$ is the total charge enclosed by the zero point space field

Where:
$\epsilon 0_{M}$ defines how much memory wave flux can pass through a memory record.
$\epsilon 0_{S}$ defines how much space wave flux can pass through a memory wave field.
$\epsilon 0_{N}$ defines how much neurowave flux can pass through a vacuum.

3. Neural Network

The voltage passing through a neural network is proportional to the enclosing neurowave field charge

$$V = k \oint_{\partial\Sigma} E \cdot dA = k\, Q_N$$

Where:

V is the voltage passing through the neural network.

k is a proportionality constant. (1)

$\oint\partial\Sigma$ is the closed surface surrounding the neural network. (2)

E is the neurowave field.

dA is a differential area element on the surface $\partial\Sigma$ (3)

Q_N is the total neurowave field charge enclosed by the surface.

(1) A proportionality constant is a factor that ensures two quantities in an equation are dimensionally consistent.

2) A closed surface integral ($\oint$) is a calculus tool used to sum up a quantity over a closed surface.

3) A differential area element (*da*) is a calculus tool used to break down complex surfaces into manageable, infinitesimally small pieces for integration purposes.

IONICS 7 SPECULATION

SPECULATION

This section Illustrates several estimates of human cognitive capacity as measured by awareness level and beingness volume.

It poses provocative questions about how dormant awareness is encapsulated and suggests retrieval strategies.

DENSITY	STATE
1	
2	
3	
4	
5	
6	
7	
8	
9	
10	
11	
12	
13	
14	
15	
16	
17	
18	
19	
20	
21	
22	
23	
24	
25	
26	
27	
28	
29	
30	
31	
32	
33	
34	
35	
36	
37	
38	
39	
40	
41	
42	
43	
44	
45	
46	
47	
48	
49	
50	
51	
52	
53	
54	
55	
56	
57	
58	
59	
60	
61	
62	
63	
64	
65	
66	
67	
68	
69	
70	
71	
72	
73	
74	
75	
76	
77	
78	
79	
80	
81	
82	
83	
84	
85	
86	
87	
88	WAKING
89	
90	
91	
92	
93	
94	
95	SLEEP
96	
97	
98	
99	

Consciousness

Treating awareness quantitatively allows us to classify states by their oton percentage, potentially enabling the construction of a periodic table of awareness.

Unconsciousness

Periodic Table of Awareness

Row	A	B	C	D	E	Level
1	O 0 Original State	I 1 Identity	2	3	4	Higher States
2 to 16	Outside Body States 5 to 79					
17	Ex 80 External Awareness	S 81 Spatial Awareness	EA 82 Energy Awareness	OA 83 Objects Awareness	84	Super-Conscious States
18	RA 85 Reflective Awareness	C 86 Cognition	E 87 Emotion	D 88 Doing	T 89 Thinking	Conscious States
19	90	91	Dy 92 Daydreaming	SL 93 Light Sleep	Dm 94 Dreaming	Subconscious States
20	Ds 95 Dreamless Sleep	96	M 97 Memory	A 98 Autonomic Nervous System	Am 99 Autonomic Memory	Unconscious States
	Being	Space	Energy	Matter	Mind	

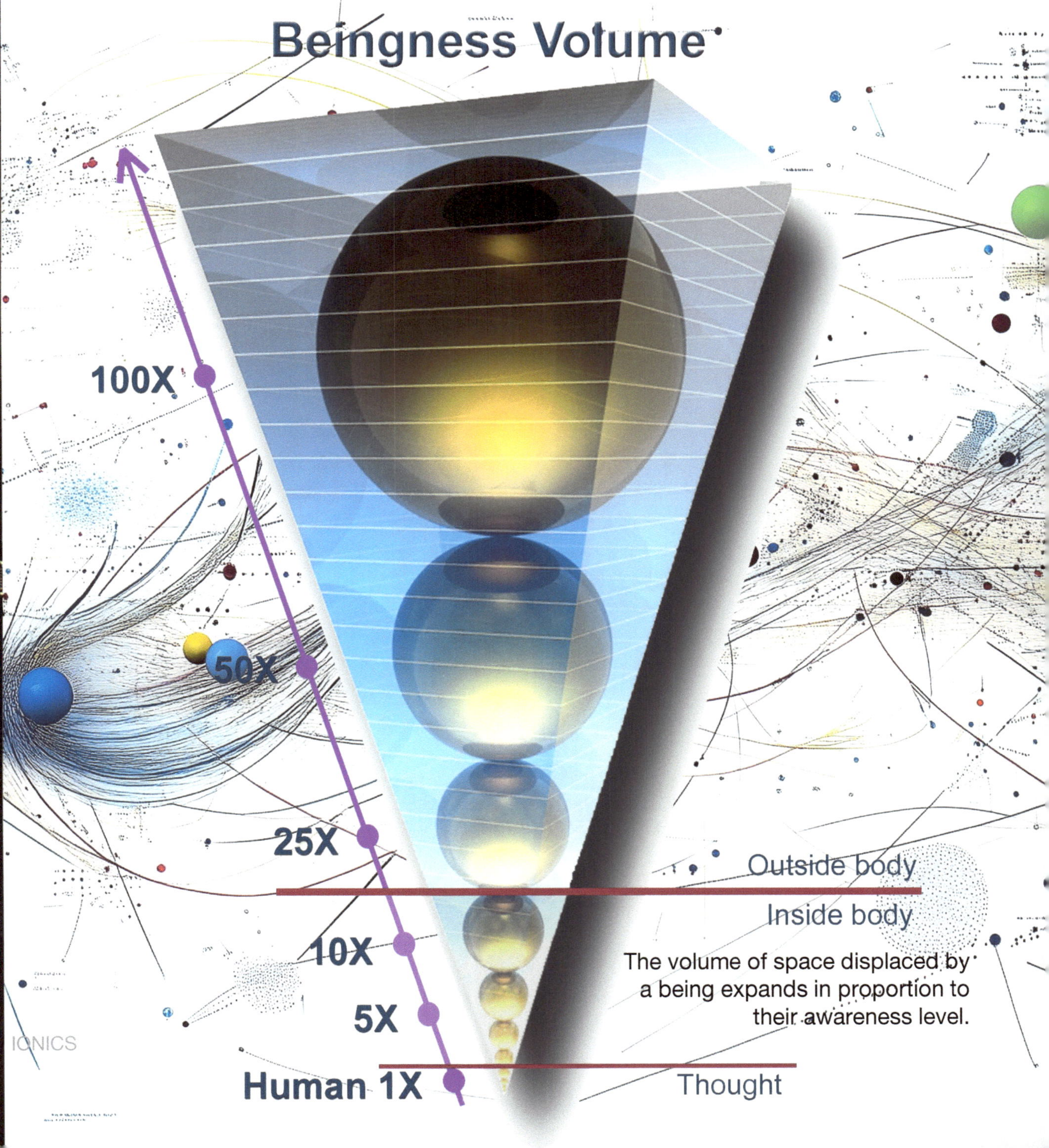

The volume of space displaced by a being expands in proportion to their awareness level.

Latent Potential

Awareness Metrics

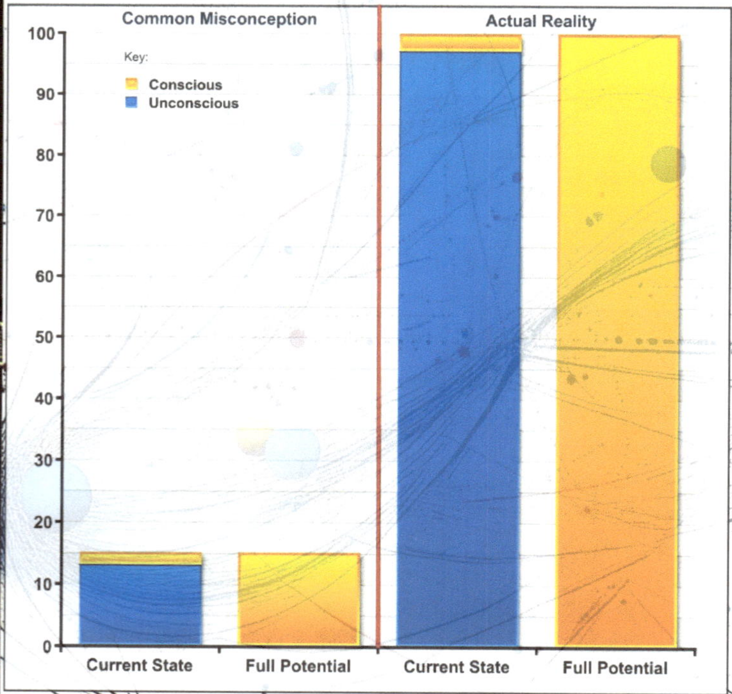

Question: If we have such enormous potential for awareness, where is it? Where is all this extra awareness hiding? Why isn't everyone superaware?

Answer: The extra awareness is unconscious.
The latent awareness we don't know we have is being soaked up by extra unconsciousness we also don't know we have.

The extra awareness is much larger than people realize (gold), and the body of unconsciousness is much greater than people think (blue).

Question: What comprises the body of unconsciousness?

Answer: Charged memories that are actively drawing otons.

Question: Children have much fewer memories than adults, yet they are not superaware. Moreover, adults have nowhere near enough memories to soak up 100X awareness.

Answer: One theory is the unconscious is primarily past life memory.

Question: Past lives?

Answer: The idea that human beings have had a large number of past lives is central in Vedanta and is elaborated in the Bhagavad Gita, Upanishads, and Brahma Sutras.

IONICS

8 REFERENCE

Glossary

io	awareness (spirit)
ioton	a particle of io
iton	free awareness particle
oton	applied awareness particle
ion	an awareness being

Units of Measure

odin	memory record charge
radin	memory wave amplitude
fallon	space wave amplitude
nanovolt	neurowave amplitude
nanoamp	neural electric current

Nomenclature

iodyne	life force
iopening	awakening
bion	embodied ion (human)
ioff	unconscious awareness
bioff	sleep
ioio	awareness-of-awareness
iout	external ion
ionova	peak hio experience
hio	higher awareness
hion	superaware ion
ionic	relating to awareness

IONICS AXIOMS

Axiom 1	Axiom 2	Axiom 3	Axiom 4
Awareness is a primary universal element.	**Awareness consists of packets.**	**Iotons have two states: free and applied.**	**Free awareness is conscious and formless.**
Awareness is fundamental and irreducible. The mind is not the brain. Awareness envelopes the brain	Io is composed of discrete packets called ***iotons,*** which are similar in principle to photons which comprise light.	Free iotons make up conscious awareness. They are called ***itons***.	Free awareness, comprised of conscious itons, is changeless, motionless and stationary.
The axioms use the term "io" for awareness. **i** representing conscious awareness **o** signifying unconscious awareness. Symbol: **ϕ** (phi)	Visualizing awareness in packets allows us to treat it numerically, opening the way to a mathematics of consciousness.	Applied iotons comprise unconscious awareness. They are called ***otons.***	The Vedic Upanishads describe this formless awareness as having no mass or energy, or location in space or time.
		An individual's awareness is a mixture of itons and otons.	**Corollary 1** **Itons have no amplitude or wavelength.** Amplitude and frequency are forms which exist in space and time. Free awareness has a non-

IONICS AXIOMS

Axiom 5	Axiom 6	Axiom 7	Formula 1
Applied awareness is unconscious and has a form.	**Law of Conservation of Awareness**	**The Mind / Matter Interface**	**Law of Conservation of Awareness**
Applied awareness, composed of otons, is located in time and oscillates in time.	Awareness cannot be created or destroyed; only changed from one form to another.	The interface between awareness and neurology comprises five layers:	$$I = (n - O)$$
Itons change state to otons when they enter a form such as a memory. The applied otons permeate and animate the form, and in this process, become unconscious.	The volumes of consciousness and unconsciousness are reciprocals of each other; as one increases, the other decreases. However, an individual's total amount of awareness remains constant.	1. Free awareness (being) 2. Information (knowing) 3. Space zero-point fields 4. Neurowave fields 5. Neural network activity	where: I = conscious awareness itons O = unconscious awareness otons n = total iotons per individual

APPLICATIONS SPECTRUM - COGNITION

COGNITIVE ENHANCEMENT

CONSCIOUS AWARENESS

- Metacognition
- Mindfulness
- Attention
- Presence
- Clarity
- Concentration
- Focus

MENTAL TRAITS

- Imagination
- Creativity
- ESP
- Memory
- Intuition
- Mental Acuity
- Reduce Mind-Wandering
- Reduce Inattention

EMOTIONAL TRAITS

- Optimism
- Emotional Balance
- Inter-connection
- Compassion
- Empathy

APPLICATIONS SPECTRUM – MENTAL HEALTH

MENTAL HEALTH

PSYCHOLOGICAL CONDITIONS

- Reduce Unwanted Behaviors
- Increase Desired Behaviors
- Lower Craving
- Psychosomatic Issues

NEUROLOGICAL CONDITIONS

- ADD/ADHD
- Stress
- OCD
- PTSD
- Anxiety
- Depression
- Insomnia
- Memory

PSYCHIATRIC CONDITIONS

- Personality Disorders
- Schizophrenia
- Bipolar

COGNITIVE IMPAIRMENT

- Alzheimer's (Early)
- Mind Cognitive Impairment
- Other

References

(1) Meditation experience is associated with differences in default mode network activity and connectivity
Judson A. Brewer, et al
Proceedings of the National Academy of Sciences USA 2011 Dec 13; 108(50): 20254–20259.
http://www.ncbi.nlm.nih.gov/pmc/articles/PMC3250176/

(2) Mystical Experiences Occasioned by the Hallucinogen Psilocybin Lead to Increases in the Personality Domain of Openness
Katherine A. MacLean, et al
Journal of Psychopharmacology 2011 Nov; 25(11): 1453–1461.
http://www.ncbi.nlm.nih.gov/pmc/articles/PMC3537171/

(3) Neural correlates of the psychedelic state as determined by fMRI studies with psilocybin
Robin L. Carhart-Harris, et al
Proceedings of the National Academy of Sciences USA 2012 Feb 7; 109(6): 2138–2143
http://www.ncbi.nlm.nih.gov/pmc/articles/PMC3277566/

(4) Psilocybin-induced spiritual experiences and insightfulness are associated with synchronization of neuronal oscillations
Michael Kometer, et al
Psychopharmacology (Berlin). 2015 Oct;232(19):3663-76.
http://www.ncbi.nlm.nih.gov/pubmed/26231498

(5) Broadband Cortical Desynchronization Underlies the Human Psychedelic State
Suresh D. Muthukumaraswamy, et al
Journal of Neuroscience, 18 September 2013, 33(38): 15171-15183;
http://www.jneurosci.org/content/33/38/15171.full

(6) Topographic pharmaco-EEG mapping of the effects of the South American psychoactive beverage ayahuasca in healthy volunteers
Jordi Riba, et al
British Journal of Clinical Pharmacology 2002 Jun; 53(6): 613–628.
http://www.ncbi.nlm.nih.gov/pmc/articles/PMC1874340/

(7) The effects of acutely administered 3,4-methylenedioxymethamphetamine on spontaneous brain function in healthy volunteers
Robin Carhart-Harris, et al
International Journal of Neuropsychopharmacology 2013 10.1017/S1461145713001405
Biological Psychiatry 2015 Oct 15;78(8):554-62
http://www.ncbi.nlm.nih.gov/pubmed/24495461

(8) The entropic brain: a theory of conscious states informed by neuroimaging research with psychedelic drugs
Robin L. Carhart-Harris, et al
Frontiers in Human Neuroscience 2014; 8: 20.
http://www.ncbi.nlm.nih.gov/pmc/articles/PMC3909994/

(9) Electroencephalogram signatures of loss and recovery of consciousness from propofol
Patrick Purdon, et al
Proceedings of the National Academy of Sciences vol. 110 no. 12, E1142-E1151
http://www.pnas.org/content/110/12/E1142.full

(10) Thalamocortical model for a propofol-induced alpha-rhythm associated with loss of consciousness
ShiNung Ching, et al
Proceedings of the National Academy of Sciences vol. 107 no. 52, 22665-22670
http://www.pnas.org/content/107/52/22665.abstract

(11) Electroencephalographic Correlates of Vasovagal Syncope Induced by Head-Up Tilt Testing
Fabrizio Ammirati, et al
Stroke 1998 Nov;29(11):2347-51
http://www.ncbi.nlm.nih.gov/pubmed/9804646

(12) Changes in EEG During Graded Exercise on a Recumbent Cycle Ergometer
Stephen P. Bailey, et al
Journal of Sports Science and Medicine (2008) 7, 505 - 511
http://www.ncbi.nlm.nih.gov/pmc/articles/PMC3761919/

(13) Changes in brain cortical activity measured by EEG are related to individual exercise preferences.
S. Schneider, et al
Journal of Physiological Behavior 2009 Oct 19;98(4):447-52.
http://www.ncbi.nlm.nih.gov/pubmed/19643120

(14) Meditation-related activations are modulated by practices needed to obtain it
Barbara Tomosino, et al
Frontiers in Human Neuroscience v.6; 2012 PMC 3539725
http://www.ncbi.nlm.nih.gov/pmc/articles/PMC3539725/

(15) Contribution of dopamine D1 and D2 receptors to amygdala activity in humans
Hidehiko Takahashi, et al
The Journal of Neuroscience, February 24, 2010 • 30(8):3043–3047

(16) Mining the posterior cingulate: segregation between memory and pain components.
Nielsen FA, Balslev D, Hansen LK.
Neuroimage: 2005 Sep; 27(3): 520-32.
https://www.ncbi.nlm.nih.gov/pubmed/15946864

(17) Neurofeedback from the Posterior Cingulate Cortex as a Mental Mirror for Meditation.
Remko van Lutterveld and Judson Brewer
Biofeedback: Fall 2015, Vol. 43, No. 3, pp. 117-120.
http://dx.doi.org/10.5298/1081-5937-43.3.05

(18) Effortless awareness: using real time neurofeedback to investigate correlates of posterior cingulate cortex activity in meditators' self-report
Kathleen A. Garrison, et al 2013
Frontiers in Human Neuroscience v.7; 2013 Aug 6 PMCID: PMC3734786
https://www.ncbi.nlm.nih.gov/pmc/articles/PMC3734786/

(19) Meditation leads to reduced default mode network activity beyond an active task
Kathleen A. Garrison, et al 2015
Cognitive, Affective, & Behavioral Neuroscience
September 2015, Volume 15, Issue 3, pp 712–720
https://www.ncbi.nlm.nih.gov/pmc/articles/PMC4529365/

(20) Dynamical Properties of BOLD Activity from the Ventral Posteromedial Cortex Associated with Meditation and Attentional Skills
Giuseppe Pagnoni, 2012
Journal of Neuroscience 11 April 2012, 32 (15) 5242-5249;
http://www.jneurosci.org/content/32/15/5242.full

(21) Impact of mindfulness on the neural responses to emotional pictures in experienced and beginner meditators
Véronique A. Taylor et al, 2011
NeuroImage, Volume 57, Issue 4, 15 August 2011, Pages 1524-1533
http://www.sciencedirect.com/science/article/pii/S1053811911006070

(22) What about the "Self" is Processed in the Posterior Cingulate Cortex?
Judson A. Brewer, Kathleen A. Garrison, and Susan Whitfield-Gabrieli, 2013
Frontiers of Human Neuroscience, 2013 Oct. 2; 7:647
https://www.ncbi.nlm.nih.gov/pmc/articles/PMC3788347/

(23) Source-space EEG neurofeedback links subjective experience with brain activity during effortless awareness meditation
Remko van Lutterveld et al, 2017
Neuroimage, 2017 May 1;151:117-127. doi: 10.1016/
https://www.ncbi.nlm.nih.gov/pubmed/26921712

(24) Diffusion markers of dendritic density and arborization in gray matter predict differences in intelligence
Erhan Genc, Christoph Fraenz, et al
Nature Communications, 2018, volume 9, Article number 1905
doi:10.1038/s41467-018-04268-8 /
http://www.nature.com/articles/s41467-018-04268-8
https://www.ncbi.nlm.nih.gov/pmc/articles/PMC5954098/

(25) Cortical glucose metabolic-rate correlates of abstract reasoning and attention studied with positron emission tomography.
Haier RJ, et al,
Intelligence. 1988; 12:199–217.
doi: 10.1016/0160-2896(88)90016-5 /
https://doi.org/10.1016/0160-2896(88)90016-5

(26) Two dose investigation of the 5-HT-agonist psilocybin on relative and global cerebral blood flow
Candace R. Lewis, Katrin H.Preller, et al
NeuroImage Volume 159, 1 October 2017, Pages 70-78
https://doi.org/10.1016/j.neuroimage.2017.07.020

(27) Neural Correlates of Personalized Spiritual Experiences, Lisa Miller, Iris Balodis et al, *Cerebral Cortex*, May 29, 2018, https://doi.org/10.1093/cercor/bhy102

(28) *Psychedelic Information Theory,* James L. Kent, Amazon, 2010, Chapter 8.

(29) University of North Carolina UNC Health Talk: This is LSD Attached to a Brain Cell Serotonin Receptor, January 26, 2017.
https://healthtalk.unchealthcare.org/this-is-lsd-attached-to-a-brain-cell-serotonin-receptor/ Article cites:

(30) Cell Press (January 26, 2017). *"Structure of LSD and its receptor explains its potency," ScienceDaily.*
https://www.sciencedaily.com/releases/2017/01/170126132541.htm

(31) Wacker D, Wang S, McCorvy JD, Betz RM, Venkatakrishnan AJ, Levit A, et al. (January 2017). *"Crystal Structure of an LSD-Bound Human Serotonin Receptor". Cell.* 168 (3): 377–389.e12. doi:10.1016/j.cell.2016.12.033. PMC 5289311. PMID 28129538. https://www.ncbi.nlm.nih.gov/pmc/articles/PMC5289311/

(32) *Lysergic acid diethylamide*, University of Bristol Chemistry Department article. http://www.chm.bris.ac.uk/motm/serotonin/LSD.HTM

(33) *Lysergic acid diethylamide*, Wikipedia article. https://en.wikipedia.org/wiki/Lysergic_acid_diethylamide

(34) Nichols DE (February 2004). "Hallucinogens". Pharmacology & Therapeutics. 101 (2): 131–81. doi:10.1016/j.pharmthera.2003.11.002. PMID 14761703. Archived from the original on February 4, 2016.

(36) Genetic modulation of the HTR2A gene reduces anxiety-related behavior in mice,
Troy T Rohn, et al, PNAS Nexus, Volume 2, Issue 6, June 2023, pgad170,
https://doi.org/10.1093/pnasnexus/pgad170

(36) Progress in Brain Research, Chapter 1 - The causal influence of conscious engagement on photonic behavior: A review of the mind-matter interaction.
Teodora Milojević, Mark A. Elliott
Elsevier, Volume 280, 2023, Pages 1-16
ISSN 0079-6123, ISBN 9780443236136,
https://doi.org/10.1016/bs.pbr.2023.03.005.

(37) Rohn, T.T., Radin, D., Brandmeyer, T. et al. Intranasal delivery of shRNA to knockdown the 5HT-2A receptor enhances memory and alleviates anxiety.
Translational Psychiatry 14, 154 (2024).
https://doi.org/10.1038/s41398-024-02879-y

(38) Electric Fields of the Brain
Paul Nunez
Oxford University Press, 2006
– circular causality
– neurowaves can move ions pages 523-4

(39) Supernormal
Dean Radin
Deepak Chopra Books, 2013
– non-local consciousness and bimodal causality

(40) Cognitive Genetics
Peter Seidler
Arcturus Press, 2024

The view that consciousness is a fundamental, irreducible aspect of the universe, distinct from the brain, is a perspective shared by various scientists and thinkers, often intersecting with the fields of parapsychology, quantum physics, and philosophy of mind. Listed below are a dozen scientists who have made significant contributions to the study of consciousness, blending science with broader philosophical and existential questions.

Dean Radin, PhD is a prominent figure in the field of parapsychology. He serves as the Chief Scientist at the Institute of Noetic Sciences and is a Distinguished Professor at the California Institute of Integral Studies. Radin has conducted extensive research on psychic phenomena, contributing significantly to the scientific study of human consciousness and extrasensory experiences. He has authored several popular books on psychic phenomena, including "The Conscious Universe" and “Supernormal.”

Rupert Sheldrake, PhD is a British biologist and author, best known for his hypothesis of morphic resonance, a concept which posits that "memory" is inherent in nature. This controversial theory challenges traditional scientific assumptions about the nature of life and consciousness. Sheldrake's work spans various subjects, including biology, parapsychology, and philosophy of science. He has written several books, such as "The Science Delusion" (published as "Science Set Free" in the US) and "Morphic Resonance".

Fred Alan Wolf, PhD is an American theoretical physicist and writer specializing in quantum physics and the relationship between physics and consciousness. He is a popular author in the field of quantum mechanics and its relation to consciousness studies, with books like "Taking the Quantum Leap" and "The Yoga of Time Travel".

Deepak Chopra, MD is an Indian-American author, public speaker, and alternative medicine advocate. A former physician, Chopra is a prominent figure in the New Age movement, known for his advocacy of mind-body spiritual healing. His approach combines principles from Ayurveda (traditional Indian medicine) with quantum physics. He has written numerous best-selling books, including "The Seven Spiritual Laws of Success" and "Ageless Body, Timeless Mind".

Edward F. Kelly, PhD is a professor of research at the University of Virginia in the Division of Perceptual Studies, which focuses on the study of consciousness and its relationship to the physical world. Kelly's work involves parapsychology, philosophy, and the mind-body problem. He has co-authored books like "Irreducible Mind: Toward a Psychology for the 21st Century" and "Beyond Physicalism", which explore the frontiers of consciousness research.

Amit Goswami, PhD is a theoretical quantum physicist and a retired professor from the University of Oregon, where he served in the Department of Physics. He is known for his work in quantum physics and its application to consciousness, spirituality, and the integration of science and spirituality. Goswami has authored several books on these topics, including "The Self-Aware Universe" and "Quantum Creativity“ and “Physics of the Soul.”

Ervin László is a Hungarian philosopher of science, systems theorist, and integral theorist. A former professor of philosophy, systems science, and futures studies, László is the founder and president of the Club of Budapest. He has authored numerous books and papers on systems theory, quantum theory, and cosmology, emphasizing the importance of consciousness in understanding the universe. His work includes "Science and the Akashic Field: An Integral Theory of Everything".

Bruce Lipton, PhD is an American developmental biologist best known for promoting the idea that genes and DNA can be manipulated by a person's beliefs. He suggests that epigenetics controls health and that the mind can play a significant role in influencing biology. Lipton's most famous book is "The Biology of Belief".

Gary E. Schwartz, PhD is a professor of psychology, medicine, neurology, psychiatry, and surgery at the University of Arizona, and the director of the Laboratory for Advances in Consciousness and Health. Schwartz's research includes parapsychology and the survival of consciousness after death. He has written several books, including "The Afterlife Experiments" and "The Sacred Promise".

Brian Josephson, PhD is a Welsh theoretical physicist and emeritus professor at the University of Cambridge. He was awarded the Nobel Prize in Physics in 1973 for his prediction of properties of a supercurrent through a tunnel barrier, known as the Josephson effect. Josephson has also delved into the study of consciousness and its relationship with quantum physics.

Larry Dossey, MD is an American physician and author known for his advocacy of the role of spirituality and mind in healthcare. He has written extensively on the non-local nature of consciousness and the healing power of prayer, with books like "Healing Words" and "One Mind".

Russell Targ is an American physicist and author, and a pioneer in the development of the laser and laser applications. He has also conducted extensive research into psychic phenomena and remote viewing. His books include "Limitless Mind: A Guide to Remote Viewing and Transformation of Consciousness".

About The Author

The physical universe and all the life within it is a creation which is floating in a sea of awareness. This Field extends infinitely in all directions, and It is self-aware. To behold Its majesty directly, one must emerge from the universe and float within the Field itself.

On the evening of July 13, 1969, Jay Mee left the physical universe and became a bubble of awareness floating in a sea of awareness. He was pure awareness inside the bubble, and pure awareness surrounded him outside the bubble. The bubble membrane was his identity as an individual being.

The ocean of awareness around him reached outwards endlessly in all directions, and it was self-aware. This was Divine Intelligence, the body of God. The physical universe was far below. It looked like a large, sparkling diamond against a backdrop of black velvet.

Although a number of people have managed to get outside of the physical universe, few have ever returned. While in the Field, Mee received cardinal insights about the nature of consciousness and the mind/matter interface which shaped his life and work.

IONICS
INSTITUTE
ionicsinstitute.org

www.ingramcontent.com/pod-product-compliance
Lightning Source LLC
LaVergne TN
LVHW071238210926
848486LV00007B/294
9781892654472